Optical Design of Microscopes

Tutorial Texts Series

- *Analysis and Evaluation of Sampled Imaging Systems*, Richard H. Vollmerhausen, Donald Reago, Ronald Driggers, Vol. TT87
- *Nanotechnology: A Crash Course*, Raúl J. Martín-Palma and Akhlesh Lakhtakia, Vol. TT86
- *Direct Detection LADAR Systems*, Richard Richmond, Stephen Cain, Vol. TT85
- *Optical Design: Applying the Fundamentals*, Max J. Riedl, Vol. TT84
- *Infrared Optics and Zoom Lenses, Second Edition*, Allen Mann, Vol. TT83
- *Optical Engineering Fundamentals, Second Edition*, Bruce H. Walker, Vol. TT82
- *Fundamentals of Polarimetric Remote Sensing*, John Schott, Vol. TT81
- *Fundamentals of Photonics*, Chandra Roychoudhuri, Vol. TT79
- *Radiation Thermometry: Fundamentals and Applications in the Petrochemical Industry*, Peter Saunders, Vol. TT78
- *Matrix Methods for Optical Layout*, Gerhard Kloos, Vol. TT77
- *Fundamentals of Infrared Detector Materials*, Michael A. Kinch, Vol. TT76
- *Practical Applications of Infrared Thermal Sensing and Imaging Equipment, Third Edition*, Herbert Kaplan, Vol. TT75
- *Bioluminescence for Food and Environmental Microbiological Safety*, Lubov Y. Brovko, Vol. TT74
- *Introduction to Image Stabilization*, Scott W. Teare, Sergio R. Restaino, Vol. TT73
- *Logic-based Nonlinear Image Processing*, Stephen Marshall, Vol. TT72
- *The Physics and Engineering of Solid State Lasers*, Yehoshua Kalisky, Vol. TT71
- *Thermal Infrared Characterization of Ground Targets and Backgrounds, Second Edition*, Pieter A. Jacobs, Vol. TT70
- *Introduction to Confocal Fluorescence Microscopy*, Michiel Müller, Vol. TT69
- *Artificial Neural Networks: An Introduction*, Kevin L. Priddy and Paul E. Keller, Vol. TT68
- *Basics of Code Division Multiple Access (CDMA)*, Raghuveer Rao and Sohail Dianat, Vol. TT67
- *Optical Imaging in Projection Microlithography*, Alfred Kwok-Kit Wong, Vol. TT66
- *Metrics for High-Quality Specular Surfaces*, Lionel R. Baker, Vol. TT65
- *Field Mathematics for Electromagnetics, Photonics, and Materials Science*, Bernard Maxum, Vol. TT64
- *High-Fidelity Medical Imaging Displays*, Aldo Badano, Michael J. Flynn, and Jerzy Kanicki, Vol. TT63
- *Diffractive Optics–Design, Fabrication, and Test*, Donald C. O'Shea, Thomas J. Suleski, Alan D. Kathman, and Dennis W. Prather, Vol. TT62
- *Fourier-Transform Spectroscopy Instrumentation Engineering*, Vidi Saptari, Vol. TT61
- *The Power- and Energy-Handling Capability of Optical Materials, Components, and Systems*, Roger M. Wood, Vol. TT60
- *Hands-on Morphological Image Processing*, Edward R. Dougherty, Roberto A. Lotufo, Vol. TT59
- *Integrated Optomechanical Analysis*, Keith B. Doyle, Victor L. Genberg, Gregory J. Michels, Vol. TT58
- *Thin-Film Design: Modulated Thickness and Other Stopband Design Methods*, Bruce Perilloux, Vol. TT57
- *Optische Grundlagen für Infrarotsysteme*, Max J. Riedl, Vol. TT56
- *An Engineering Introduction to Biotechnology*, J. Patrick Fitch, Vol. TT55
- *Image Performance in CRT Displays*, Kenneth Compton, Vol. TT54
- *Introduction to Laser Diode-Pumped Solid State Lasers*, Richard Scheps, Vol. TT53
- *Modulation Transfer Function in Optical and Electro-Optical Systems*, Glenn D. Boreman, Vol. TT52
- *Uncooled Thermal Imaging Arrays, Systems, and Applications*, Paul W. Kruse, Vol. TT51
- *Fundamentals of Antennas*, Christos G. Christodoulou and Parveen Wahid, Vol. TT50
- *Basics of Spectroscopy*, David W. Ball, Vol. TT49
- *Optical Design Fundamentals for Infrared Systems, Second Edition*, Max J. Riedl, Vol. TT48
- *Resolution Enhancement Techniques in Optical Lithography*, Alfred Kwok-Kit Wong, Vol. TT47
- *Copper Interconnect Technology*, Christoph Steinbrüchel and Barry L. Chin, Vol. TT46
- *Optical Design for Visual Systems*, Bruce H. Walker, Vol. TT45

Optical Design of Microscopes

George H. Seward

Tutorial Texts in Optical Engineering
Volume TT88

SPIE
PRESS
Bellingham, Washington USA

Library of Congress Cataloging-in-Publication Data

Seward, George.
Optical design of microscopes / George Seward.
p. cm. -- (Tutorial texts in optical engineering ; v. TT88)
ISBN 978-0-8194-8095-8
1. Microscopes. 2. Optics. I. Title.

QH211.S478 2010
681'.413--dc22

2010000660

Published by

SPIE
P.O. Box 10
Bellingham, Washington 98227-0010 USA
Phone: +1 360 676 3290
Fax: +1 360 647 1445
Email: Books@spie.org
www.spie.org

Printed in the United States of America.

Introduction to the Series

Since its inception in 1989, the Tutorial Texts (TT) series has grown to more than 80 titles covering many diverse fields of science and engineering. The initial idea for the series was to make material presented in SPIE short courses available to those who could not attend and to provide a reference text for those who could. Thus, many of the texts in this series are generated by augmenting course notes with descriptive text that further illuminates the subject. In this way, the TT becomes an excellent stand-alone reference that finds a much wider audience than only short course attendees.

Tutorial Texts have grown in popularity and in the scope of material covered since 1989. They no longer necessarily stem from short courses; rather, they are often generated by experts in the field. They are popular because they provide a ready reference to those wishing to learn about emerging technologies or the latest information within their field. The topics within the series have grown from the initial areas of geometrical optics, optical detectors, and image processing to include the emerging fields of nanotechnology, biomedical optics, fiber optics, and laser technologies. Authors contributing to the TT series are instructed to provide introductory material so that those new to the field may use the book as a starting point to get a basic grasp of the material. It is hoped that some readers may develop sufficient interest to take a short course by the author or pursue further research in more advanced books to delve deeper into the subject.

The books in this series are distinguished from other technical monographs and textbooks in the way in which the material is presented. In keeping with the tutorial nature of the series, there is an emphasis on the use of graphical and illustrative material to better elucidate basic and advanced concepts. There is also heavy use of tabular reference data and numerous examples to further explain the concepts presented. The publishing time for the books is kept to a minimum so that the books will be as timely and up-to-date as possible. Furthermore, these introductory books are competitively priced compared to more traditional books on the same subject.

When a proposal for a text is received, each proposal is evaluated to determine the relevance of the proposed topic. This initial reviewing process has been very helpful to authors in identifying, early in the writing process, the need for additional material or other changes in approach that would serve to strengthen the text. Once a manuscript is completed, it is peer reviewed to ensure that chapters communicate accurately the essential ingredients of the science and technologies under discussion.

It is my goal to maintain the style and quality of books in the series and to further expand the topic areas to include new emerging fields as they become of interest to our reading audience.

James A. Harrington
Rutgers University

Contents

Chapter 4 Aberrations / 27

Chapter 5 Basic Physical Optics / 35

Chapter 6 Fluorescence / 53

Chapter 7 Optical Design Metrics / 63

Chapter 8 Image Contrast / 73

Chapter 9 Microlens Formats / 85

Chapter 10 Illumination Systems / 93

Chapter 11 Cover Strata / 105

Chapter 12 Objective Lenses / 113

Chapter 13 Tube Elements / 127

Chapter 14 Ocular Lenses / 139

Chapter 15 Sensors / 147

Chapter 16 Human Vision / 157

Preface

This book provides an introduction to optical design as it pertains to microscopes. The large numerical aperture of a microscope creates issues that are not present in systems such as telescopes and cameras. The importance of microscope design is growing rapidly in 2010.

Microscopes are frequently employed in drug development, clinical tests, and genomics; however, one system does not fit all applications. Considerable expertise is required for the evaluation, design, and manufacture of these instruments. An integrated relationship must be established between several subsystems: the source, the illumination optics, the specimen, the objective lens, the tube optics, and the sensor.

In this text, the optical prescriptions are specified with glass name, refractive index, and Abbe number. Methods for color correction are described with specific materials. The anomalous partial dispersion in the blue is described for a long crown and a short flint. The origin of dispersion is related to glass composition.

As this is a tutorial, several essential topics in optics are reviewed. Geometric optics provides a simplistic description of ray tracing. Physical optics is described in two chapters: the first chapter describes the basic concepts of wave propagation with simple algebra, and the second chapter describes more advanced concepts with vector calculus. Optical aberrations describe an optical system's departure from perfection. Expressions for image contrast are defined for practical application to imaging systems. Expertise in these topics is essential for proper design of an optical instrument.

Numerous chapters rely on simple plots and basic algebra, so even a novice designer should easily comprehend the majority of chapters; however, a few chapters require comprehension of calculus, vector operations, and Fourier analysis. An expert in design should seek maximum comprehension of the chapter on advanced concepts at the end of the book.

This book provides a foundation for the development of expertise in optical design. Expertise is developed through education, practice, and exploration.

George H. Seward
March 2010

Chapter 1
Optical Design Concepts

1.1 A Value Proposition

A comprehensive optical design should yield both consistent manufacture and maximum performance. Such features are critical to biological applications, where the cost of consumables can be far greater than the cost of the optical hardware. Maximum performance can reduce the magnitude of consumables on a daily basis. Furthermore, a false negative can delay proper treatment of a critical condition. Consistent manufacture builds credibility in a clinical test.

Sections 1.2 through 1.9 provide a review of essential optical design concepts. Expertise within these concepts is developed through education and practice. Perseverance is not a substitute for relevant expertise in optical design.

1.2 Specimen Model

The structure of a specimen determines the limits of optical performance. A small deviation from the nominal thickness can introduce a significant wavefront error that is not corrected by a lens. The light collected by the margin of the lens is not focused within a diffraction-limited spot. The marginal light is spread over many pixels, which lowers image contrast. Consequently, an optical design must accommodate the tolerances of the specimen. In particular, the cover strata of biological specimens present significant challenges in optical design.

1.3 Detector Parameters

Every detector has fundamental limitations, which may be expressed as equivalent incident photons. A dark current is converted to photons per time per area. Both full-well capacity and read noise are converted to photons per area. The dependence on area is very important because image contrast is frequently determined by image incidence (photons per second per area). Thus, an equivalent incidence of the detector must be established.

1.4 Numerical Aperture

The margin of a lens is defined by the aperture of the lens. The angular extent of the aperture varies with immersion medium. The numerical aperture (NA) defines a constant throughout all media. The marginal NA is defined as

$$\mathrm{NA}_M = n\sin\theta_M, \tag{1.1}$$

where n is the refractive index and θ_M is the angle of the marginal ray.[i] The marginal ray is defined as a ray from the object point to the margin of the aperture. The NA can decrease with field height, yielding less image brightness at the field margin.

The marginal NA defines several important features of the lens system. The spot diameter [Eq. (1.4)] depends on NA_M^{-1}. The depth of focus depends on NA_M^{-2} [Eq. (1.8)]. The collection efficiency by solid angle roughly depends on NA_M^{2} [Eq. (8.11)]. The peak irradiance of a diffraction-limited spot depends on NA_M^{4} [Eq. (8.24)]. The NA is an essential metric for an optical instrument.

The paraxial NA is derived from the focal length f and entrance pupil diameter ϕ_{Ent}:

$$\mathrm{NA}_P = \frac{\phi_{Ent}}{2f} = \frac{1}{2F}, \tag{1.2}$$

where F is the *f*-number (*f*/#). The principal surface is a sphere with its center at the focal point. The marginal NA can differ from the paraxial NA because spherical aberration can bend marginal rays both too much or too little. The *f*/# can inflate with field height, which yields less image brightness. The image-space *f*/# may also inflate as the focal length grows with field position.

1.5 Wave Propagation

The space-angle product describes the convergence of a spherical wavefront as[1]

$$A\Omega \geq \lambda^2, \tag{1.3}$$

where A is the area of point spread, Ω is the solid angle of convergence, and λ is the wavelength. The diffraction limit of the point spread is defined by the minimum space-angle product.

A planar version of the space-angle product is

$$d\,\mathrm{NA}_G = 1.27\lambda, \tag{1.4}$$

where d is the diameter of the Gaussian beam and NA_G is the Gaussian NA. The area of the Gaussian point spread doubles over the Rayleigh distance, which is

[i] The term numerical aperture has the unique distinction of being a two-letter symbol that persists in contrast to the more common practice of using a one-letter symbol in math and physics.

$$z_R = \frac{A_0}{\lambda} = \frac{\lambda}{\Omega}. \tag{1.5}$$

A doubling of the point-spread area also indicates a halving of the point-spread irradiance. Consequently, the Rayleigh distance indicates a depth of focus with a defined range of contrast.

The space-angle product for a circular aperture is

$$d\,\mathrm{NA}_M = 2.44\lambda, \tag{1.6}$$

where d is the diameter of the Airy disc.

A Gaussian estimate of an Airy point spread is

$$\mathrm{NA}_G = 0.71\,\mathrm{NA}_M. \tag{1.7}$$

The Gaussian depth of focus may be expressed by marginal NA as

$$z_R = 2\frac{\lambda}{\pi\,\mathrm{NA}_M^2}. \tag{1.8}$$

This relationship provides an effective metric for depth of focus in relation to wave propagation.

1.6 Geometric Aberrations

Geometric aberrations describe the departure of convergent rays from an image point. As the geometric point spread grows beyond the diffraction limit, the actual point spread becomes aberration limited. An aberration-limited spot frequently contains a sharp central peak and a broad annular skirt. The central peak may be defined by the central NA (NA_C), which is smaller than the marginal NA. Thus, there are four NAs for an aberration-limited lens: NA_G, NA_C, NA_M, and NA_P. Their relationships by size are expressed as follows:

$$\mathrm{NA}_G < \mathrm{NA}_C \le \mathrm{NA}_M \approx \mathrm{NA}_P. \tag{1.9}$$

Geometric aberrations are described by polynomials within a ray-intercept plot: a cubic plot indicates spherical aberration; a parabolic plot indicates coma; dissimilar slopes indicate astigmatism or axial color; a vertical shift indicates lateral color. Comprehension of ray-intercept plots is essential for evaluation of optical designs.

Production optics can be effectively modeled through patent prescriptions. Specifications within patents should resemble the actual or a similar product. A proper optical design tool can generate ray-intercept plots, wavefront errors, point spreads, and line spreads. A practical field height may be derived from the onset of coma and astigmatism. Chromatic aberrations are common in both patent prescriptions and actual products. A central NA should be derived from the metrics of the optical design tool. Collection efficiency is derived from an encircled-energy plot or calculated from the central and marginal NAs.

A production lens can also be modeled from published modulation transfer function (MTF) data. The onset of astigmatism is frequently displayed in MTF data, and a central NA may be estimated from the shape of the MTF data.

1.7 Image Contrast

Ordinarily, increasing the paraxial NA improves image contrast by directing more object light into a smaller point spread; however, this is not always true. Increasing the paraxial NA can degrade image contrast by directing more object light into the annular skirt of an aberration-limited point spread.

A contrast calculation is based on several features. The central NA determines the collection efficiency of the object, whereas the marginal NA determines the collection efficiency of the background. The size of an object determines the appropriate calculation of collection efficiency by a pixel. A full-pixel object has a 100% collection efficiency by pixel, while a subpixel object does not. A fluorophore radiates from a point, while intrinsic fluorescence radiates from a full pixel. These are important nuances in the detection of fluorophores.

The resolution of an optical system may be defined in numerous formats. An Airy disk is a frequently used metric in optical design. A geometric radius may indicate a point spread far beyond the Airy radius. A root-mean-square (rms) radius provides a practical metric for resolution. An ensquared energy plot effectively defines collection efficiency of a pixel. A central NA defines a sharp peak within the point spread. A wavefront error describes the limitations of plano optics, such as filters and mirrors.

The specimen flux (quanta per time) must be considered. A fluorophore might stop emitting after emission of 100 photons. Consequently, the collection NA must be maximized. However, increasing the collection NA does not always increase the image contrast.

The background radiance must also be considered. It can originate from the foreground or background, or can even originate from the optical path. Impurities in the cover strata can emit fluorescence, as can the intrinsic material of the lens.

Detector noise is yet another consideration. Dark current, read noise, and dark noise are three common metrics of electronic detectors of light. The object signal may overcome the read noise through integration; however, the object signal cannot overcome the dark current by integration. The dark noise is the sum of the dark current and read noise at a specific frame rate. The shot noise of the object signal must also be considered.

There are numerous possibilities for the definition of contrast. An effective model requires careful consideration of the application. It is important to note that collection of more light does not always improve image contrast; an optimum NA yields the maximum contrast.

1.8 Manufacturing

Surface quality is expressed in three common formats. Surface-form error describes the deviation from the specified form; 50 nm is a typical specification for surface-form error in optics for the visible spectrum. Surface imperfections describe the size and number of digs, coating defects, and scratches (known as "scratch/dig"). Surface imperfections must be extremely small when near the object field. Surface texture defines the fine structure limits. The total integrated scatter (TIS) of a surface is defined as

$$\mathrm{TIS} = \phi^2, \tag{1.10}$$

where ϕ is the rms phase error of the exiting wavefront.

The material quality is extremely important. Stress birefringence creates wavefront aberration, whereas bubbles, inclusions, and striae cause scattering. Impurities create extrinsic fluorescence and may discolor during ultraviolet (UV) exposure. Crystalline inclusions may be dragged through the surface during polish. Some vendors provide metrics for abrasion resistance. Material quality is often overlooked during fabrication specifications. The cost of optical glass increases rapidly as the purity increases; however, a thrifty choice of glass often results in more expense due to rejections during system assembly.

The key to consistent manufacturing is specification of reasonable tolerances. A reasonable tolerance is based on the current art of production. Such limitations are not normally published data. They must be discovered through discussion with the vendor. An acceptable lens design might fail in production due to unreasonable tolerances.

1.9 Assembly

Assembly procedures should state specific steps toward a measurable result. Otherwise, performance may vary between units. Tools may be required for qualification of subsystems, and subsystems should be qualified for installation into larger assemblies. A facile assembly procedure is essential for consistent production. After assembly, the system must be qualified. A standard specimen should be specified for measurement by the instrument and may require significant expense.

Chapter 2
Basic Microscope Concepts

2.1 Magnification

Magnification specifies the ratio of an image dimension to its object dimension. Magnification is typically cited in one of three formats: axial magnification, radial magnification, or angular magnification. Axial magnification defines the ratio of an axial image shift to an axial object shift. Radial magnification defines the ratio of an image height to the object height. Axial magnification is proportional to the square of the radial magnification. Angular magnification defines the ratio of the image angle to the object height or angle.

Several types of dimensions are employed in defining these magnifications: an axial dimension is a distance along the optical axis; a radial dimension is a distance across the optical axis, and may be linear or angular. A radial dimension may be called lateral or transverse, and an axial dimension may be called longitudinal.

In a microscope, the "magnification" can be both radial and angular. The objective lens indicates a radial magnification for a real image, while an ocular lens specifies an angular magnification for a virtual image. A microscope does not typically specify an axial magnification, although it is an important parameter, since the depth of focus is scaled by the axial magnification.

2.2 Accommodation

The human eye comprises a static lens of the cornea and a dynamic lens within the ciliary muscle. The standard far point of human vision is infinity ∞, which may be expressed as zero diopters of magnification. A standard eye should focus on an object at 250 mm. This represents an addition of four diopters by the compressed dynamic lens. Therefore, the standard near point s_{NP} of human vision is 250 mm, or four diopters of magnification. The process of dynamic focus within this range (250 mm–∞) is called accommodation.

The standard optical power of the near point is defined as

$$\phi_{NP} = \frac{1}{s_{NP}} = 40 \text{ diopters}, \tag{2.1}$$

where the optical distance of the standard near point is 250 mm. A diopter is a reciprocal meter.

2.3 Finite Tube Length

Until the late 1900s, most microscopes employed a lone objective lens to create an image at the end of a tube. The objective flange mates to the front end of a tube, and an eyepiece mounts to the back end of the tube. The tube length defines the nominal operation of the objective. Standard tube lengths vary by vendor from 160 to 210 mm. An objective lens is designed specifically for a nominal tube length, which is normally marked on the objective barrel. An objective lens with finite tube length does not function properly at another tube length.

Figure 2.1 displays a thin-lens model of an objective lens with a finite tube length. The lateral magnification is negative; therefore, the image is inverted. It is upside down and backward with respect to the object. Consequently, the image moves in the opposite direction of the object. The objective lens may be called "the objective" without reference to a lens.

The absolute magnification M of an objective lens with a finite tube length is fairly estimated by Eq. 3.7 as

$$(1+M) \approx \frac{L_T}{f_{Obj}}, \tag{2.2}$$

where L_T is the length of the tube, and f_{Obj} is the focal length of the objective lens. An objective barrel is normally marked with both tube length and magnification. The magnification M drops the negative polarity of the radial magnification. The prescription of an objective is tailored to a particular magnification. Insertion of plano optics within the convergent rays of the tube can create significant aberration and defocus.

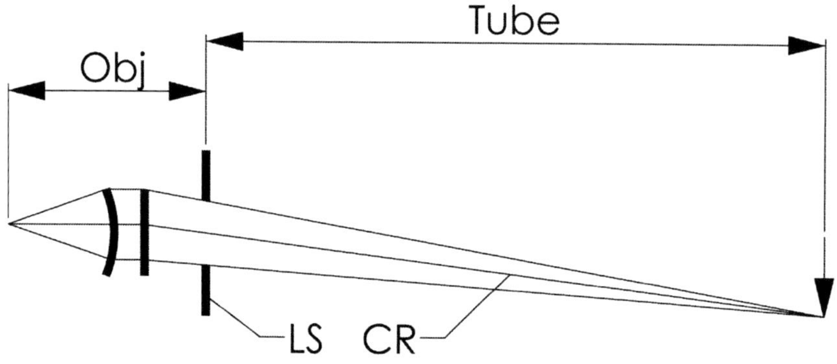

Figure 2.1 Objective lens with a finite tube length. The tube defines a finite distance to the image conjugate. The chief ray (CR) travels from the center of the lens stop (LS) to the margin of the image.

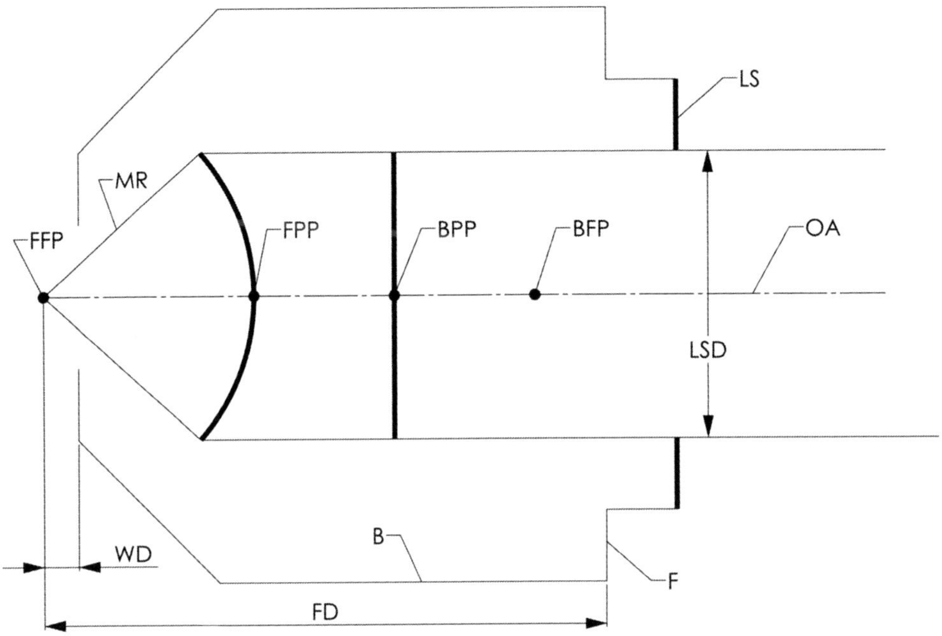

Figure 2.2 Objective lens with infinity correction. The cardinal points along the optical axis (OA) are front focal point (FFP), front principal plane (FPP), back principal plane (BPP), and back focal point (BFP). The lens stop defines the marginal ray (MR) and lens-stop diameter (LSD). Body (B) defines working distance (WD), flange (F), and flange distance (FD).

2.4 Infinity-Corrected Objective

An ∞-corrected microscope splits the objective lens into two parts: an objective lens and a tube lens. The objective lens creates an infinitely distant image within the tube. The tube lens converts the infinitely distant image into a real image at the tube-lens focal point. Plano optics may be placed in the tube without creation of aberration.

Figure 2.2 displays a modern objective lens with ∞ correction. The object is located at the front focal point. The image conjugate is infinitely distant. The objective prescription is specified for an infinitely distant image. The ∞ correction is frequently indicated on the objective barrel as ∞, or INF. The flange distance specifies the nominal distance of the object from a flange on the barrel. The working distance indicates a gap between the nominal object position and the body of the microscope.

The paraxial NA of an objective lens defines the angular size of the lens stop with respect to the field:

$$\mathrm{NA}_P = n \sin\theta_{LS} = \frac{\phi_{LS}}{2f}, \tag{2.3}$$

where θ_{LS} is the angle of the lens stop, ϕ_{LS} is the diameter of the lens stop, and f is the focal length. The marginal ray is defined by the center of the object field and the margin of the lens stop.

2.5 Tube Lens

A tube lens converts the infinitely distant conjugate of an objective into an image at the back focal point of the tube lens. The magnification of an ∞-corrected objective lens is specified as

$$M_{Obj} = \frac{f_T}{f_{Obj}}, \tag{2.4}$$

where f_T is the focal length of the tube lens and f_{Obj} is the focal length of the objective lens. The magnification will certainly change if the focal length of the tube lens is changed. The image at the back focal point of the tube-lens is both real and inverted.

The ∞ correction of the tube lens permits insertion of plano optics without incurring defocus or aberration. Multiple beamsplitters and filters can also be inserted without issue. However, a warped beamsplitter is not a plano optic; the optics must be flat within a specific tolerance.

The location of the lens stop determines the location of the chief ray within the tube lens. Separation of rays by field within a tube lens enables selective correction of off-axis aberrations. Typically, a tube lens may travel axially by 10% of its focal length without creating significant aberration. A smaller allowance of 1% applies to lateral movement.

Figure 2.3 displays a tube lens in combination with an objective lens. The lens stop is placed at the front focal point of the tube lens. If the lens stop is considered an object, then the tube lens creates an infinitely distant image of the lens stop. The distant lens stop defines a telecentric lens. A telecentric image space is also indicated by an axial chief ray at the image. The object space is also telecentric.

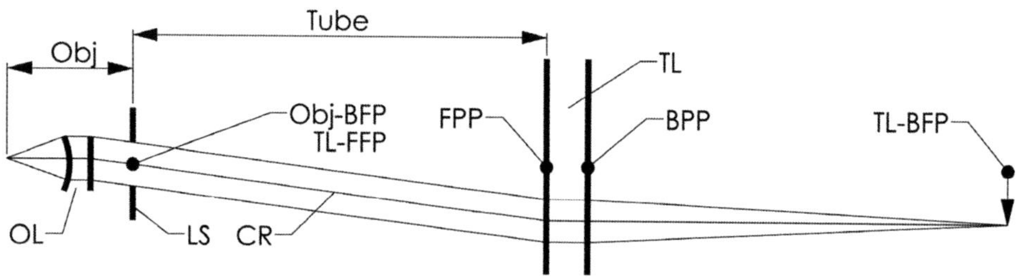

Figure 2.3 Objective lens (OL) with tube lens (TL). The objective lens has a back focal point (Obj-BFP). The tube lens has a front focal point (TL-FFP), front principal plane (TL-FPP), back principal plane (TL-BPP), and back focal point (TL-BFP). The chief ray travels from the center of the lens stop to the margin of the image.

2.6 Ocular Lens

An ocular lens is a magnifier with a lens stop on the front side. It is frequently called an eyepiece or ocular without reference to a lens. The ocular creates an image of the lens stop at the back focal point of the ocular. The exit pupil of a microscope is defined by a bright white disk floating above the ocular. This floating disk is an image of the lens stop of the objective lens. It might even be an image of a filament from the illumination system. The exit pupil of a microscope is typically smaller than the 2.5-mm natural pupil of the eye. Consequently, the eye operates within a diffraction limit when looking into a microscope. The exit pupil of a microscope can significantly sharpen visual acuity.

The diameter of the exit pupil is derived from Eq. (3.8) as

$$\phi_{Ex} = \phi_{LS} \frac{f_{Oc}}{f_{Oc} - \left(f_{Oc} + f_T\right)}, \tag{2.5}$$

where f_{Oc} is the focal length of the ocular. Simplification yields

$$\left|\phi_{Ex}\right| = \phi_{LS} \frac{f_{Oc}}{f_T}. \tag{2.6}$$

The eye relief specifies the distance from the back of the ocular to the exit pupil. Eye relief allows room for spectacles. Furthermore, an eye cup may reduce contributions from ambient light.

As displayed in Fig. 2.4, an ocular creates an image of the lens stop. This image defines an exit pupil at the back focal point of the ocular lens (see Table 2.1). The ocular lens also magnifies an image at its front focal point. The magnification of an ∞-corrected ocular lens is specified as

$$M_{Oc} = \frac{\phi_{Oc}}{\phi_{NP}} = \frac{250 \text{ mm}}{f_{Oc}}, \tag{2.7}$$

where ϕ_{Oc} is the power of the ocular, ϕ_{NP} is the power of the standard near point at 250 mm, and f_{Oc} is the focal length of the ocular lens.

Table 2.1 Two common geometries for an exit pupil.

Objective lens stop	Focal length Tube	Focal length Ocular	Exit pupil	Brand
10	200	25	1.25	Nikon
10	180	25	1.39	Olympus

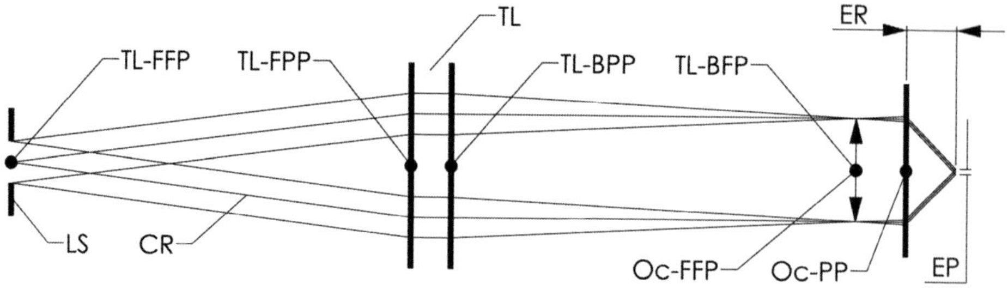

Figure 2.4 Tube lens with an ocular lens. The ocular lens has a front focal point (Oc-FFP), principal plane (Oc-PP), and back focal point. The exit pupil is located at the ocular back focal point. The exit pupil (EP) is an image of the lens stop. The eye relief (ER) defines the distance from the last surface to the exit pupil. The chief ray travels from the center of the lens stop through the margin of the image to the center of the exit pupil.

2.7 Refractive Objects

A refractive object, or refractile, deflects illumination beyond the vision lens stop. Figure 2.5 displays a thin-lens model of a cell that acts on a single point of illumination. In Fig. 2.5, the object is centered on the illumination, and the entire illumination ray bundle passes through the vision lens stop. The center of the cell appears bright white. In Fig. 2.6, the object is slightly decentered, while the entire illumination ray bundle barely passes the vision lens stop. The cell remains bright. In Fig. 2.7, the object is decentered enough for partial blocking by the vision lens stop. The cell appears gray at this point. The edge of the cell gradually grows darker in the image. In Fig. 2.8, the object is decentered enough for complete blocking by the vision lens stop. The edge of the cell remains dark over a short distance from the edge.

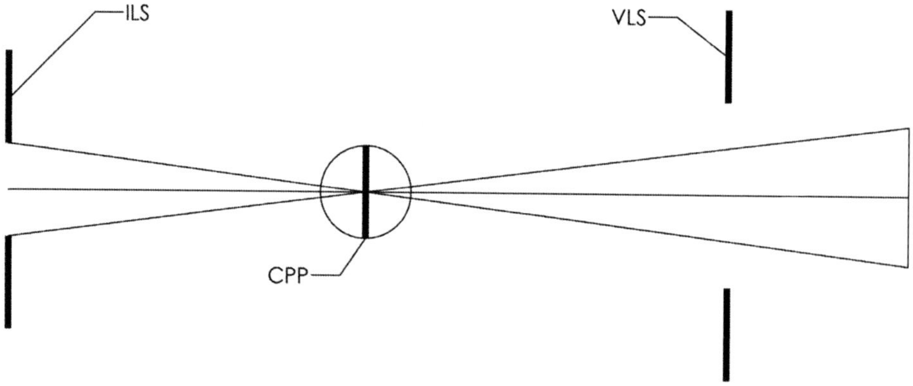

Figure 2.5 Bright field at the center of a cell. The cell image is bright in the center of the cell. The illumination lens stop (ILS) defines the convergence of illumination. The vision lens stop (VLS) defines the collection of illumination. The background appears bright.

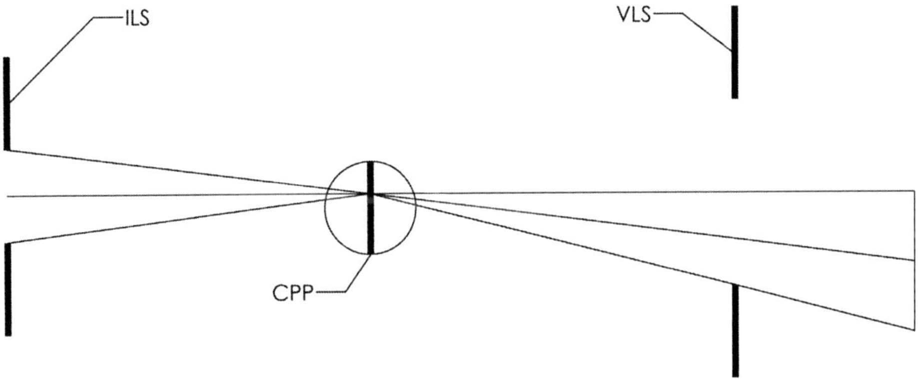

Figure 2.6 Bright field near the edge of a cell. The cell image is bright near the edge of the cell. The illumination lens stop defines the convergence of illumination. The vision lens stop defines the collection of illumination. The background appears bright.

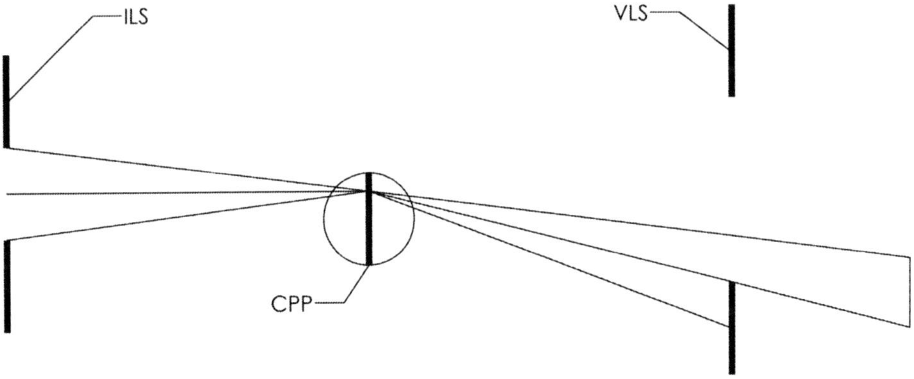

Figure 2.7 Bright field in transition at the edge of a cell. The cell image is gray during the transition to dark at the edge of the cell. Refraction by the cell directs light into the opaque portion of the lens stop.

The width of the gray portion of refraction is proportional to the illumination NA (NA_I):

$$w_{GR} \propto NA_I, \quad \text{for}\,(NA_I < NA_V)\;. \tag{2.8}$$

As NA_I deceases, the gray band decreases in width.

The width of the dark region is dependent on the vision NA (NA_V) as below:

$$w_{DR} \propto (1 - NA_V - NA_I), \text{ for}\,(NA_I < NA_V)\,. \tag{2.9}$$

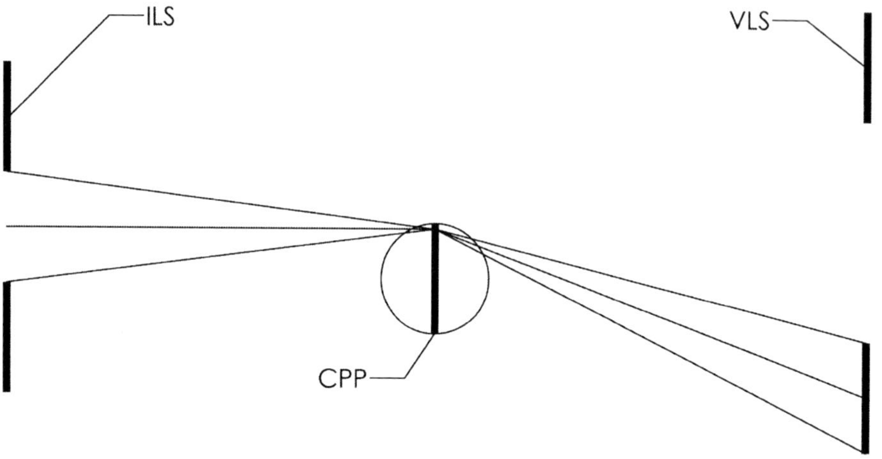

Figure 2.8 Bright field at the dark edge of a cell. The cell image is dark at the edge of the cell. Refraction by the cell directs light into the opaque portion of the lens stop.

As the vision NA grows, a steeper portion of the cell wall is required for sufficient deflection. Both of these linear dependencies are dependent on a thin-lens model. Actual dependencies can be nonlinear.

The first-order resolution of a refractive object is dependent on NA_I and NA_V. Reduction of NA_I should reduce the total width of an edge of refraction. However, the width of the dark portion also displays important dependencies on diffraction.

2.8 Diffractive Objects

A diffractive object implies some form of edge. An edge has numerous origins, such as the phase delay of an index change or the absorption by a dye. A refractive object has a diffractive component at the edge.

A circular object creates the well-known Airy pattern. The angular components of the Airy pattern tend toward zero at a larger angle; however, they are finite in magnitude up to 90 deg. Consequently, the vision lens stop defines the angular extent of the circular object. The spatial resolution of diffraction is dependent on wavelength and NA_V:

$$w_D \propto \frac{\lambda}{NA_V}. \tag{2.10}$$

2.9 Dark Field

Figures 2.9 and 2.10 display a dark field. In Fig. 2.9, the central portion of the vision lens stop blocks the normal illumination, and the image background appears dark. The center of the cell also looks dark. In Fig. 2.10, the edge of the cell refracts light into the open annulus of the vision lens stop, and the edge of the cell appears bright.

Diffraction may also create bright features in a dark field. An edge diffracts light through the lens stop with little dependence on NA_I. A subpixel object may create significant diffraction through the vision lens stop. Moreover, a dark-field contrast may reveal small features that are not visible in the bright field.

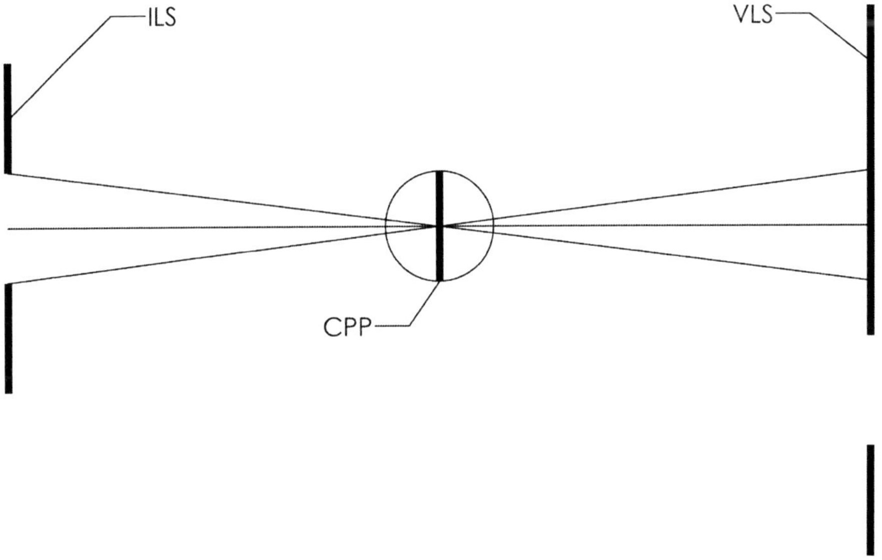

Figure 2.9 Dark field at the center of a cell. The cell image is dark at the center of the cell. The center of the vision lens stop blocks illumination in the background. The background appears dark.

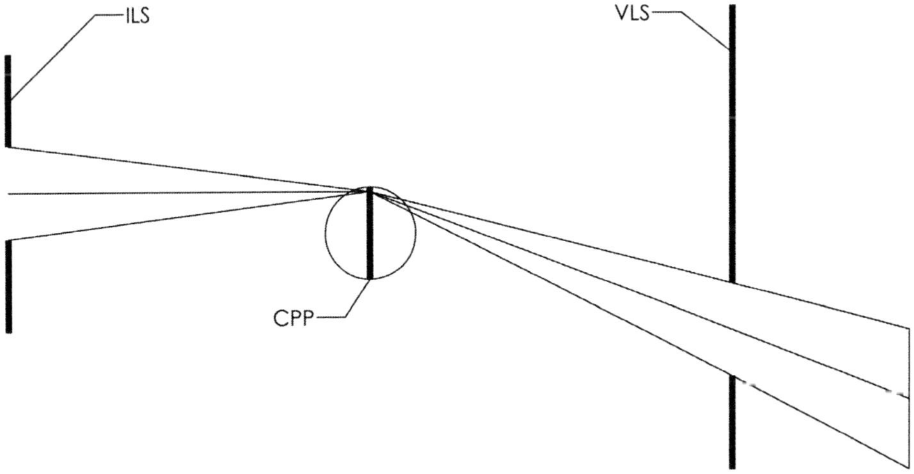

Figure 2.10 Dark field at the edge of a cell. The cell image is bright at the edge of the cell. Refraction by the cell directs light into the clear portion of the vision lens stop. The clear portion may be a partial or full annulus. A partial annulus may rotate about the optical axis.

Chapter 3
Basic Geometric Optics

3.1 Ray Tracing

A ray is a convenient representation of wave propagation. A ray represents the normal to a wavefront. Geometric optics defines the transport of light through simple geometric constructions of ray transmission and reflection. Snell's law is the only consideration during transmission of a ray. Paraxial optics specifies cardinal points without consideration of aberrations. Seidel aberrations indicate defects through polynomials within ray-intercept plots. Comprehension of the ensuing topics is essential for the effective design of an optical instrument.

It is important to remember that geometric optics is a simple model for more complex wave propagation. Ignorance or negligence of physical optics frequently yields a dysfunctional design. Ray tracing is a convenient and powerful tool; however, it is not a complete description of an optical system.

3.2 Cardinal points

There are three cardinal points in a lens: the principal point, the focal point, and the nodal point. These cardinal points are easily managed in systems with identical refractive indices at the object and image. However, the immersion oil of a high-power objective lens requires a more complex set of cardinal points.

The principal point of a lens defines an axial point of refraction by the lens. The principal point of a lens may be extended into a principal plane of the lens or even a principal surface of a spherical shape. All refraction occurs at the principal plane. A lone principal plane defines a thin lens where all rays are bent only at the principal plane. A principal plane may also be split into a front and back principal plane, which is the definition of a thick lens. There is no radial displacement of rays between the principal planes of a thick lens. A thick lens provides a simple description of a complex lens through paraxial optics.

The focal point of a lens defines a point of convergence for incident axial rays. A focal length is defined by the distance from a principal plane to a focal point. If the refractive index of the image is different from the refractive index of the object, then there are two different focal lengths: a front focal length for the incident rays of the object and a back focal length for the exiting rays of the image. The effective focal length (EFL) normally cites the focal length in air.

A nodal point of a lens defines a nodal ray that is not refracted by the lens. The exiting nodal ray travels in the same direction as the incident nodal ray.

Ordinarily, the nodal points are coincident with the principal points; however, if the refractive index of the front is different from that of the back, then the nodal points are separated from the principal points.

3.3 Stops

The aperture of a lens defines a stop, which restricts flow of light. A field stop defines the spatial extent of an image field. A lens stop defines the angular extent of the aperture at an image. Other stops might eliminate detrimental rays. The word aperture can be ambiguously applied to a variety of thick-lens parameters; therefore, it is important to reference an aperture by its proper name: field stop, lens stop, or back focal plane. The marginal ray defines the angular extent of the lens stop with respect to an image point. A chief ray travels through the margin of the field stop and the center of the lens stop; it is frequently referenced in ray calculations.

The marginal NA (NA_M) is derived from the marginal ray of the lens stop as

$$NA_M = n \sin \theta_M , \tag{3.1}$$

where n is the refractive index and θ_M is the angle of the marginal ray. The NA can be applied to several formats. The marginal ray can apply to the object, the image, the focal point, or even the angular field of view. The NA must be properly indentified by its location.

The *f*/# of a lens is another metric for stop size:

$$F = \frac{f}{\phi_{EnP}} , \tag{3.2}$$

where ϕ_{EnP} is the diameter of the entrance pupil as defined by an image of the lens stop within object space. The *f*/# of a lens depends on an image at a focal point. The *f*/# does not consistently correlate to NA, which is dependent on image distance. The image-space *f*/# of a lens is

$$F_I = \frac{s_I}{\phi_{ExP}} , \tag{3.3}$$

where ϕ_{ExP} is the diameter of the exit pupil, and s_I is the image distance. As stated earlier, the exit pupil is an image of the lens stop within image space. Conversely, the entrance pupil is an image of the lens stop within object space. As the image distance increases, the image-space *f*/# inflates, thus reducing the brightness of subpixel objects.

The image-space *f*/# can inflate for several reasons. It doubles for a relay lens with unity magnification and inflates at large field angles due to a reduction of the solid angle with field position. A lens design might require *f*/# inflation for control of aberration. Spherical aberration might inflate the effective number for subpixel objects. A low-*f*/# camera lens (such as *f*/1.4) should be carefully examined before using it for low-light level applications. The image contrast will not likely match calculations of the *f*/# stated by the vendor.

3.4 Gaussian Lens Formula

A thin-lens model describes a lens in terms of refraction at a principal plane. Figure 3.1 displays the four cardinal points of a thin-lens diagram. By convention, incident rays are governed by the front focal point, whereas exiting rays are governed by the back focal point. The front focal length is longer than the back focal length. However, the rules of the cardinal points are still observed. An incident front ray becomes a back axial ray. A nodal ray maintains slope. A front focal ray becomes a back axial ray.

The image conjugates are related by distance, focal length, and refractive index:

$$\frac{n_O}{d_O} + \frac{n_I}{d_I} = \frac{n_O}{f_O}, \tag{3.4}$$

where n_O is the refractive index within object space, n_I is the refractive index within image space, d_O is the object distance from the principal plane, d_I is the image distance from the principal plane, and f_O is the focal length within object space. This concept may also be written as the Gaussian lens formula:

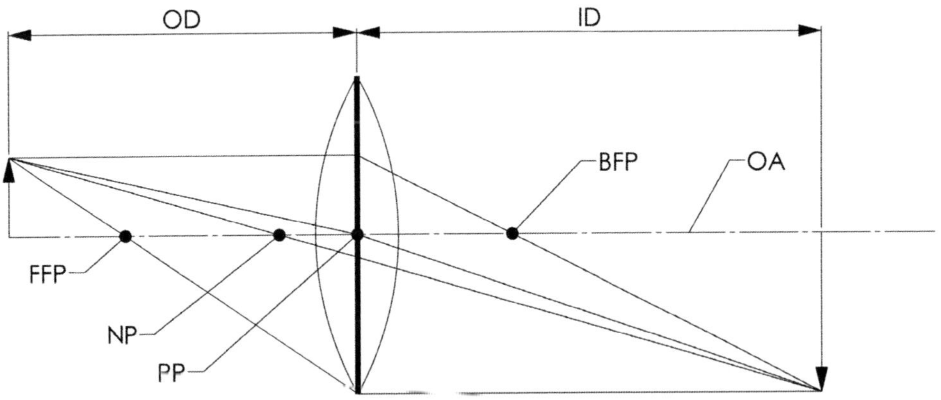

Figure 3.1 Thin-lens model of a relay lens with refractive indices of 1.5 for incident rays on the front side and 1.0 for exiting rays on the back side. Front focal point (FFP), nodal point (NP), principal point (PP), back focal point (BFP), object distance (OD), and image distance (ID) are indicated. The OD is positive to the left (from PP to the front space). The ID is positive to the right (from PP to the back space).

$$\frac{1}{s_O}+\frac{1}{s_I}=\frac{1}{f}, \tag{3.5}$$

where s_O is the optical distance of the object, s_I is the optical distance of the image, and f is the focal length within air. Optical distance is the equivalent distance in air.

A traditional Gaussian lens formula employs a specific polarity convention. A positive focal length or distance is measured from a principal plane toward a focal point. Verbal descriptions of image conjugates can be difficult. Consequently, a thin-lens diagram is warranted.

The lateral magnification M_L can be defined in several formats:

$$M_L=\frac{h_I}{h_O}=\frac{-s_I}{s_O}, \tag{3.6}$$

where h_I is the height of the image and h_O is the height of the object. In terms of focal length, the magnification is expressed as

$$(1-M_L)=\frac{s_I}{f}=\frac{-M_L s_O}{f}. \tag{3.7}$$

$$M_L=\frac{f}{f-s_O}. \tag{3.8}$$

During positive lateral magnification, the image moves in the same direction as the object. The axial magnification is

$$M_A=\frac{\Delta s_I}{\Delta s_O}=M_L^2. \tag{3.9}$$

The tilt magnification is

$$M_T=\frac{M_A}{M_L}=M_L. \tag{3.10}$$

During positive lateral magnification, the image tilts in the same direction as the object. The magnification of a lens is normally stated as the absolute value of the lateral magnification:

$$M=|M_L|. \tag{3.11}$$

The reduction of a lens is the reciprocal of the magnification:

$$R = \frac{1}{M}. \tag{3.12}$$

3.5 Image Types

Images may be considered real or virtual. A real image is formed by rays that actually pass through the image point. Consequently, a finger may touch a real image. A virtual image, on the other hand, is formed by rays that apparently originate from the image point but do not exist at the image point. A virtual image does not exist in real space. A finger cannot touch a virtual image.

Image orientation may be defined as the orientation of its upside and left side. An erect image is "upside-up" and "left-side-left." A flipped image is either upside down or backward. An inverted image is flipped in both directions—it is upside down and backward. A mirror image is flipped along just one direction. Finally, an image conjugate is an image within a pair of images conjoined by a lens system. The object may be considered an image conjugate.

3.6 Optical Power

The optical power (ϕ) of a lens is defined as the reciprocal of the focal length within air:

$$\phi = \frac{n}{f_n} = \frac{1}{f}, \tag{3.13}$$

where n is the refractive index and f_n is the focal length within the refractive index. A diopter specifies a unit for optical power as reciprocal meters. Thus, a focal length of 1000 mm indicates one diopter of power, while a focal length of 250 mm defines four diopters of power.

There are three formats of the optical power: transmission, external reflection, and internal reflection. The polarity is positive when a collimated incident beam becomes a convergent exiting beam. The optical power for transmission is

$$\phi_T = \Delta n \left(\frac{1}{R} \right), \tag{3.14}$$

wherein Δn is the change in refractive index and R is the radius of the surface. The optical power for an external reflection is

$$\phi_{ER} = 2 \left(\frac{1}{R} \right). \tag{3.15}$$

The optical power for an internal reflection is

$$\phi_{IR} = 2n\left(\frac{1}{R}\right). \tag{3.16}$$

At an air-to-glass interface (n = 1.0, 1.5), the optical power of transmission is

$$\phi_T = 0.5\left(\frac{1}{R}\right). \tag{3.17}$$

The optical power of an external reflection is

$$\phi_{ER} = 2\left(\frac{1}{R}\right) = 4\phi_T. \tag{3.18}$$

The optical power of an internal reflection is

$$\phi_{IR} = 3\left(\frac{1}{R}\right) = 6\phi_T. \tag{3.19}$$

Obviously, the effect of radius is much greater in reflection than transmission. The effects of surface texture and surface form error display a similar dependence.

Optical power may also define angular magnification. The angular height per spatial height within air may be defined through optical power as

$$\phi = \frac{n}{s}, \tag{3.20}$$

where s is spatial distance to the object. The angular height of an object is defined as

$$\alpha = \frac{h}{s} = h\frac{\phi}{n}, \tag{3.21}$$

where h is the spatial height of an object. The spatial height of an object or image is independent of the local refractive index, while the angular height is inversely proportional to the local refractive index.

3.7 Paraxial Optics

The term paraxial defines a regime where the rays are nearly equivalent to the optical axis, and the following approximation is valid:

$$\sin\theta = \theta. \tag{3.22}$$

The paraxial condition may also apply to an off-axis ray that is nearly equivalent to the normal. The paraxial condition is a first-order approximation of Snell's law:

$$n_1\theta_1 = n_2\theta_2 . \tag{3.23}$$

This linear relationship enables a simple-matrix method for the construction of images. The optical distance is the equivalent distance in air:

$$s = \frac{d}{n}, \tag{3.24}$$

where d is the spatial distance and n is the refractive index. The optical distance is expressed in units of spatial length. The optical path length is a length of the phase within a wave:

$$\Lambda = \frac{nd}{\lambda}, \tag{3.25}$$

where Λ is the optical path length and λ is the spatial wavelength. The optical path length is expressed in units of phase angle, cycles, or wavelengths. The optical path difference normally refers to the optical path length. It is important to carefully define these parameters, as they are frequently confused.

3.8 Relay Lens

Figure 3.1 displays a thin-lens model of a relay lens. The cardinal points are located in dissimilar refractive media. The object is immersed in oil at a refractive index of 1.5 on the front side of the lens. The front focal length is 1.5 times the back focal length. An image of the object is formed on the back side of the lens within air at a refractive index of 1.0. A focal ray is converted to an axial ray at the principal plane, and vice versa. A nodal ray is transmitted at the principal plane without refraction. The image location is established through two of three rays: an incident axial ray, an incident focal ray, or an incident nodal ray.

Incident rays are governed by the front cardinal points, whereas exiting rays are governed by the back cardinal points. The front focal point of a positive lens is located on the incident side of the lens, whereas the front focal point of a negative lens is located on the exiting side of the lens.

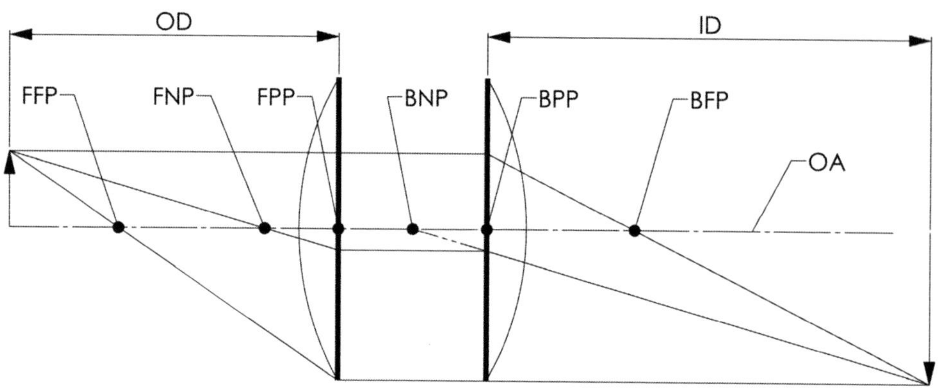

Figure 3.2 Thick-lens model of relay lens with refractive indices of 1.5 for incident rays on the front side and 1.0 for exiting rays on the back side. Front nodal point (FNP) and back nodal point (BNP) are shown.

Figure 3.2 displays a thick-lens model of a relay lens. The principal plane is split into front and a back principal planes. Rays translate from the front to back principal planes without a change in height. There are now two principal points and two nodal points. Refraction in a thick lens depends on cardinal points in the same fashion as a thin lens. A thick-lens model provides accurate location of cardinal points within a lens barrel.

3.9 Magnifier

Figure 3.3 displays a thin-lens model for a magnifier. The object is located inside the front focal point. The image is located outside of the front focal point. The real rays on the back side of the lens do not extend to the front side. The real rays are extended through the front side as virtual rays that originate from a virtual image. A virtual ray does not exist in real space.

In Fig. 3.4, the angular magnification of a magnifier is defined by an object at its front focal point. The magnifier converts the object into an infinitely distant image. The angular size of the image with respect to the principal point is

$$\alpha n = h_O \phi_M , \tag{3.26}$$

where α is the angle in radians, n is the refractive index, h_O is the object height, f is the focal length, and ϕ_M is the power of the magnifier. Thus, the angular magnification is the optical power of the magnifier. The lateral magnification of an object at the standard near point is

$$M_L = \frac{\phi_M}{\phi_{NP}} = \frac{s_{NP}}{f_M} . \tag{3.27}$$

A typical magnifier displays 40 diopters of optical power, while the near-point optical power is 4 diopters. Application of a 40-diopter magnifier to the standard near point yields a lateral magnification of 10 diopters by angle.

The loupe was an early format of a magnifier. Its name is derived from the French word for an imperfect gem; its modern translation is "magnifying glass." A modern loupe may contain one or more lens elements. Walker[2] describes several magnifiers in detail.

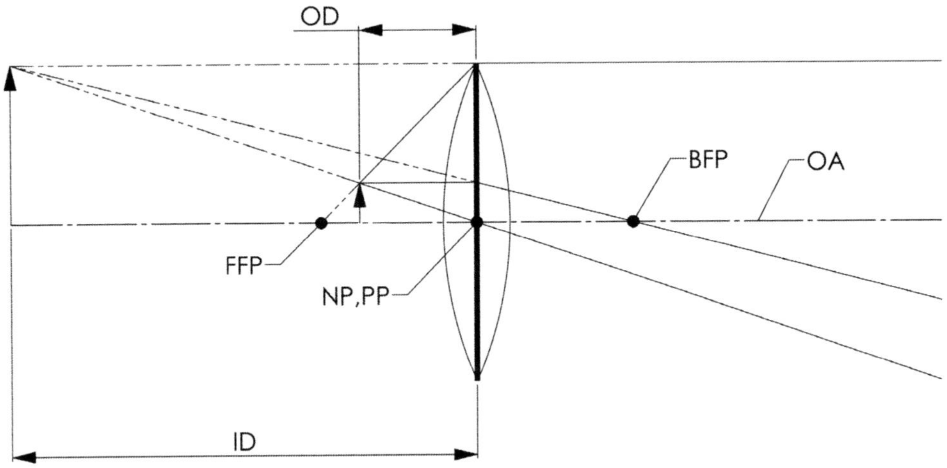

Figure 3.3 Thin-lens model of a magnifier. Phantom lines (dash, dash, long dash) indicate virtual rays and a virtual object. Solid lines indicate real rays.

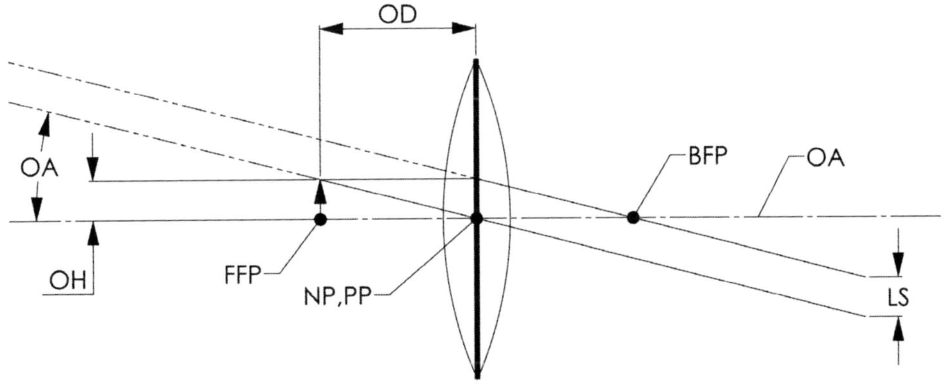

Figure 3.4 Magnification of a magnifier with the object at the front focal point. Object angle (OA) is defined by object height (OH) times magnification in diopters of power.

Chapter 4
Aberrations

As a consultant, this author has reviewed the design and performance of numerous instruments and three common problems were found: ignorance of physical optics, negligence of aberrations, and absence of a tolerance budget. At least one member of the design team should indentify the type of aberration within the lens. Spherical aberration grows with NA, coma grows with field angle and NA, and astigmatism grows with tilt of the optic. These are important considerations for an optical system design.

4.1 Seidel Aberrations

A polynomial expression for the ray height at an image is expressed as follows:[3]

$$\begin{aligned} y &= M h \\ &+ A_1 p \\ &+ B_0 p^3 + B_1 h^1 p^2 + B_2 h^2 p^1 + B_3 h^3 \\ &+ \sum\nolimits_0^5 C_i h^i p^{5-i} + \cdots, \end{aligned} \tag{4.1}$$

where h is the object height and p is the pupil position. The first line indicates the paraxial image height, which is simply the product of the magnification M and the object height h. The paraxial image height specifies an image system without aberration. The second line indicates a first-order aberration: defocus A_1. The third line indicates third-order aberrations: spherical aberration B_0, coma B_1, astigmatism and Petzval curvature B_2, and distortion B_3.

Defocus occurs in several formats: an axial error in the position of the sensor creates defocus across the entire image; a curved image surface creates variable defocus across the field; a tilted surface creates astigmatism; a variation in refractive index creates axial color. Defocus is indicated by a dependency on p^1.

Figure 4.1 displays a report for a spherical lens with spherical aberration. There is a dependency on p^3. Spherical aberration is created by an increased power of the marginal focus with respect to the axial focus. The marginal rays are bent too much due to the spherical shape of the surface.

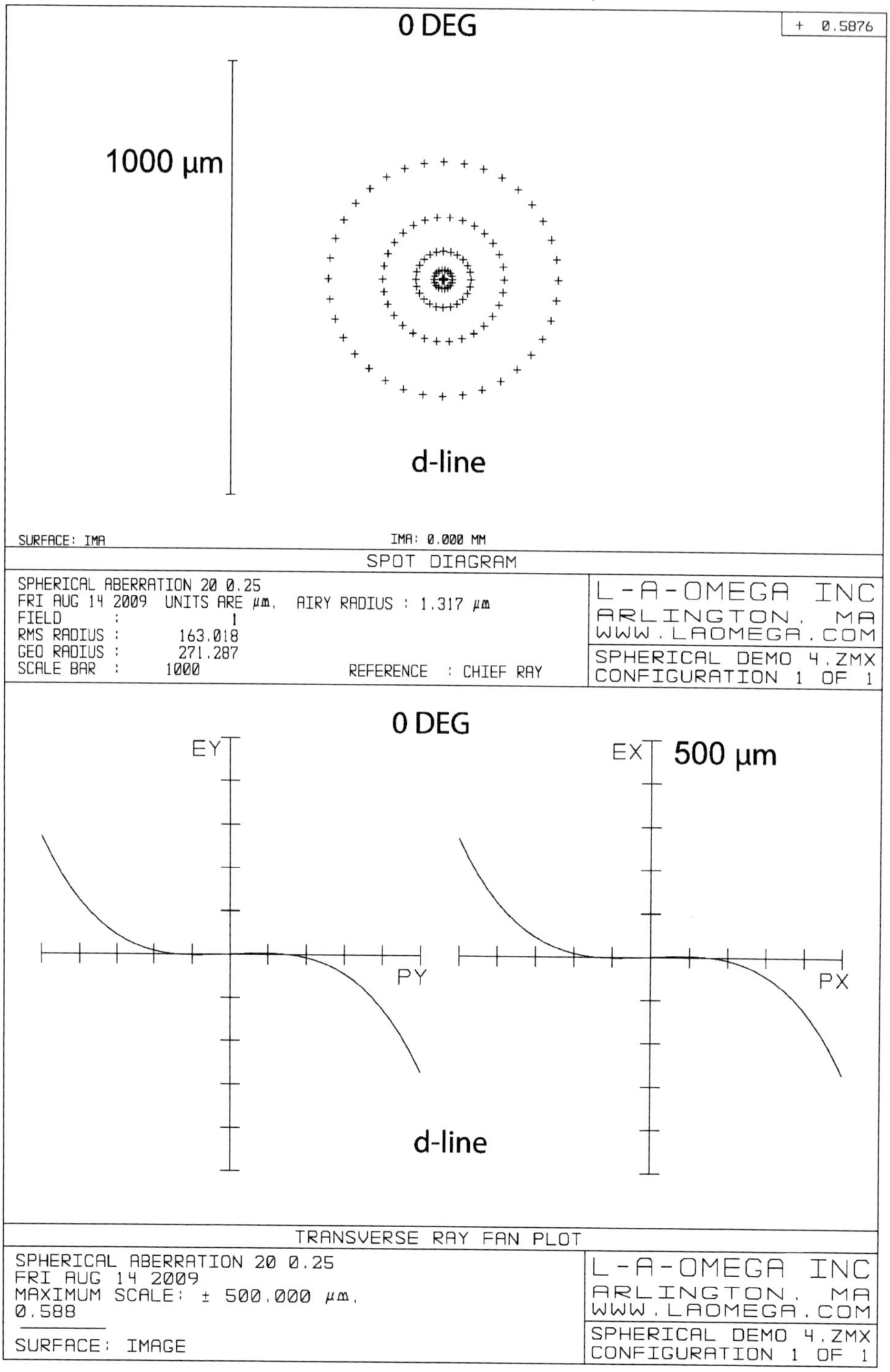

Figure 4.1 Spherical aberration ray fan and spot. The spherical lens has a 20-mm focal length, BK7 glass, and 0.25 NA.

Figure 4.2 displays a report for an aspheric lens with coma. There is a dependency on hp^2. Coma creates the appearance of hairy stars which resemble comets. Comet is derived from the Greek word *komē*, meaning hairy, and *astēr*, meaning star. Coma is created by both an axial and a lateral shift of the marginal focus.

Figure 4.3 displays a report for an aspheric lens with astigmatism. There is a dependency on h^2p, which occurs in the tangential plane. Astigmatism is derived from the Greek *a*-, indicating an antonym, and *stigma*, meaning spot. Ergo, "astigmatism" means "no spot," as in a line. There is focus along the tangential pupil[i] EY, while the spot is spread within the sagittal plane[ii] EX. Astigmatism is created by an axial shift of the tangential focus with respect to the sagittal focus. The tangential pupil covers a larger arc angle due to the tilt of the lens. Consequently, the tangential power is greater than the sagittal power.

The Petzval radius describes the radius of the paraxial focal surface. A defocus by field position depends on h^2p. Distortion describes a growth in focal length with field position. The increased image height depends on h^3, which is not normally revealed in a ray fan. Both Petzval curvature and distortion are normally described through dedicated plots for those features. For example, a grid distortion plot might reveal a barrel or pincushion distortion of a square.

4.2 Chromatic Aberrations

Figure 4.3 displays lateral color. The ray intercept of the F and C lines are dependent on h. Lateral color is created by the normal dispersion of glass, wherein the refractive index grows as the wavelength decreases. The F line (486 nm) is shifted more than the C line (656 nm).

Figure 4.4 displays a report for an aspheric lens at 0.25 NA. There is a large slope in the ray-intercept plots for all wavelengths in the x pupil but not in the y pupil. The ray intercept of the d line is essentially flat, which indicates no defocus. The ray intercept of the F line displays a negative slope, which indicates axial defocus. The F line is bent more strongly due to a higher refractive index. As with lateral color, axial color is created by the normal dispersion of glass, wherein the refractive index grows as the wavelength decreases.

4.3 Other Aberrations

Third-order aberrations indicate any sum of three for the powers of h and p. There are also fifth-order aberrations that become more important at higher NAs. Spherochromatism indicates a change in spherical aberration with wavelength.

Errors in manufacture or assembly can also create aberrations. A lateral shift between elements of equal but opposite spherical aberration can create significant coma. Surface form error can also add random error. Finally, thin optics could warp during application of coatings.

[i] The tangential plane contains the optical axis and the field point.

[ii] The sagittal plane contains the chief ray and is normal to the tangential plane.

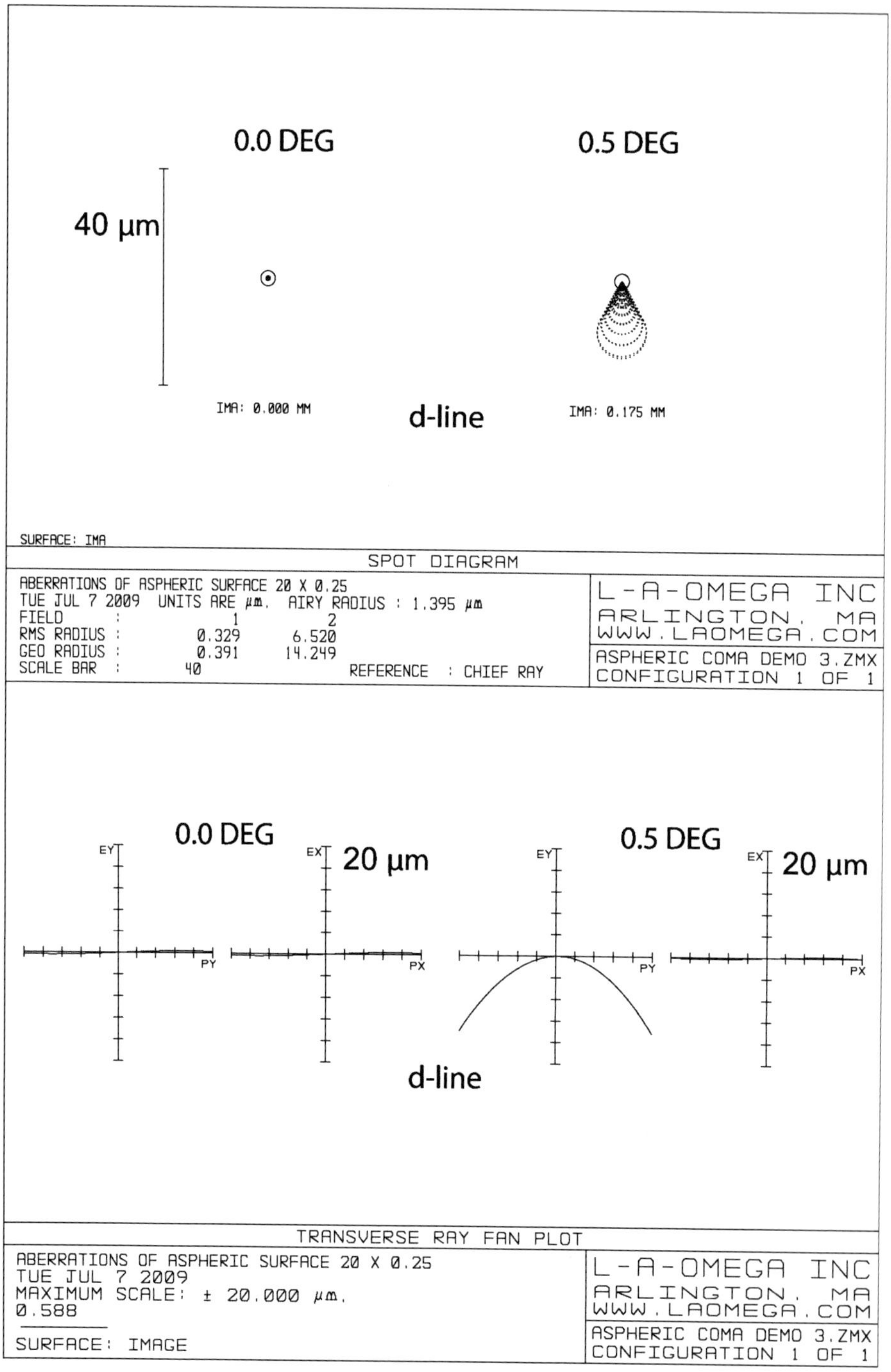

Figure 4.2 Coma ray fan and spot. The aspheric lens has a 20-mm focal length, BK7 glass, and 0.25 NA.

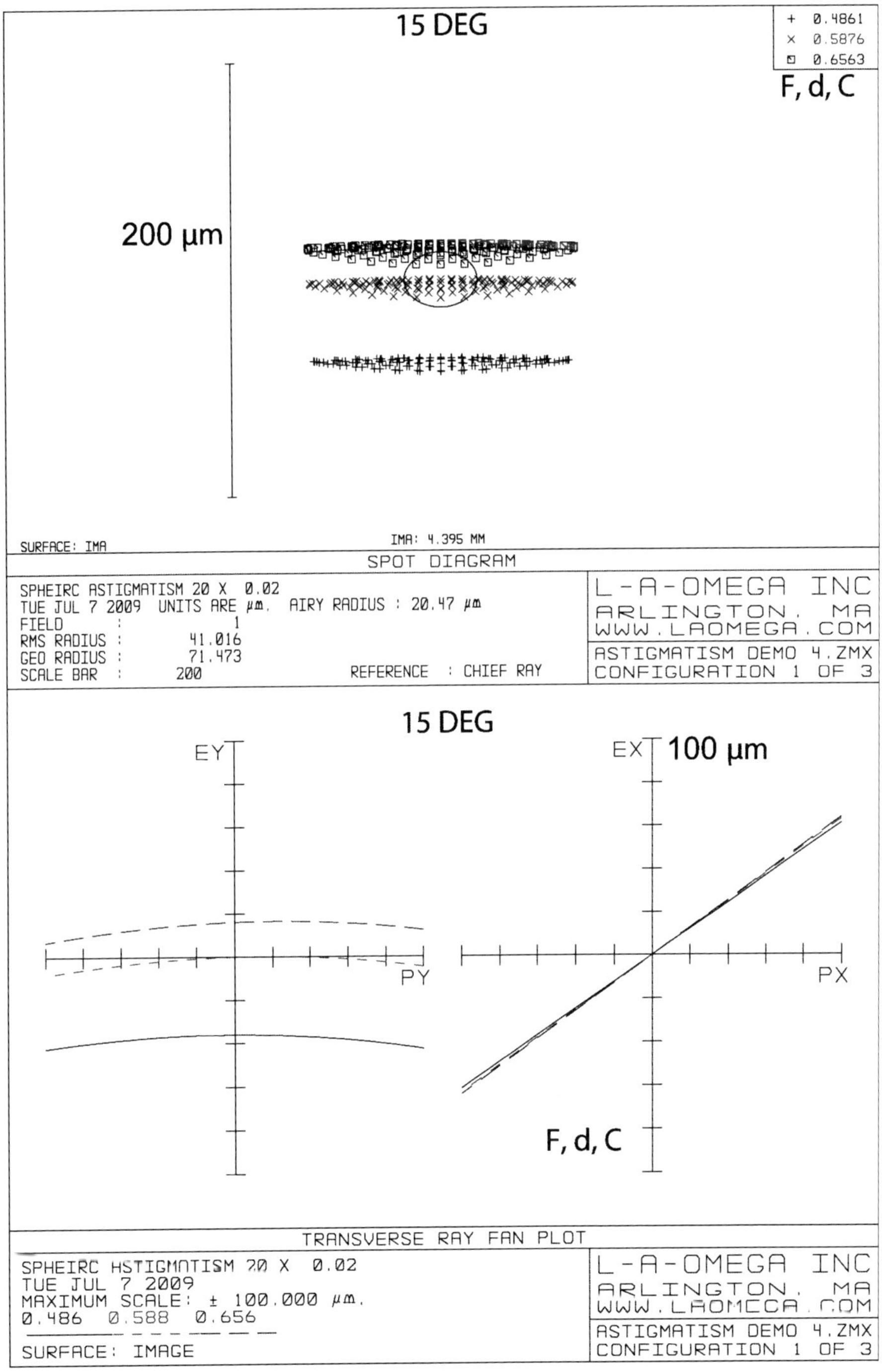

Figure 4.3 Astigmatism and lateral color ray fan and spot. The spherical lens has a 20-mm focal length, BK7 glass, and 0.02 NA.

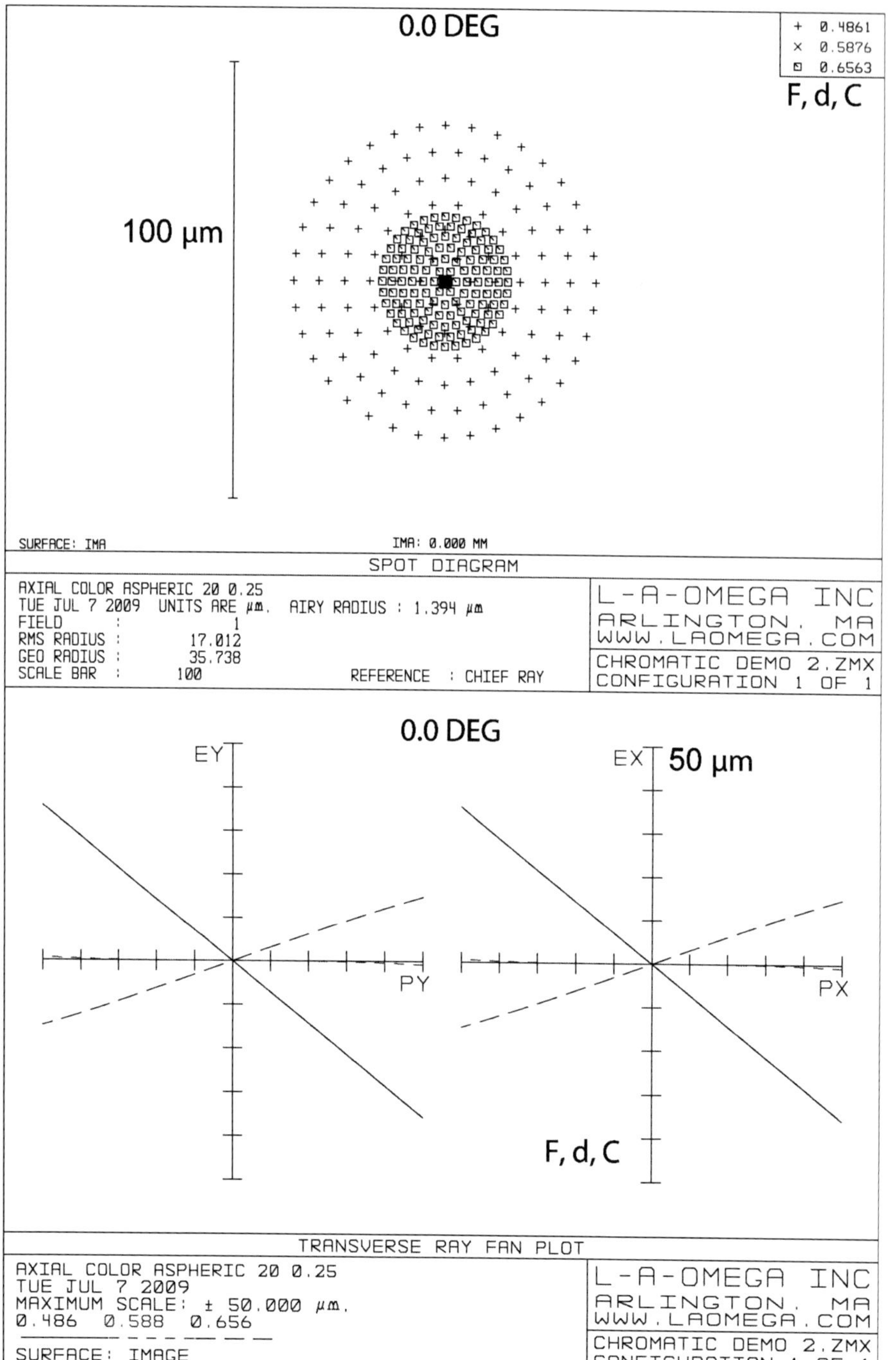

Figure 4.4 Axial color ray fan and spot. The aspheric lens has a 20-mm focal length, BK7 glass, 0.20 NA, 0-deg field, F line (486 nm), d line (588 nm), and C line (656 nm).

4.4 Aspheric Surfaces

An aspheric surface is frequently defined by its conic section plus a summation of polynomials, as below:

$$z = R\frac{(y/R)^2}{1+\sqrt{1-(1+K)(y/R)^2}} + \sum A_n y^n , \tag{4.2}$$

where R is the radius of curvature, y is the height above the z axis, and K is the conic constant. A conic constant of 0.0 defines a sphere that is easily created by polishing. A conic constant of –1.0 defines a parabola, whereby reflection converges a collimated beam into a point. A conic constant between 0.0 and –1.0 defines an ellipse, whereby internal refraction effectively converges a collimated beam into a point. Finally, a conic constant beyond –1.0 defines a hyperbola, whereby external refraction effectively converges a collimated beam into a point. The Brewster window defines a practical limit for refractive aspheric surfaces. A reflective parabola is limited only by tolerances on the field and surface form and texture.

The polynomial coefficients A_n are useful for correction of aberrations. An A_2 coefficient is normally zero because it is prominent in the conic section. A conic constant of –0.4334 at the lens stop may correct for spherical aberration at the cost of coma, as in Fig. 4.2. An A_4 near the lens stop may also correct third-order spherical aberration. An A_4 and A_6 near the field stop may correct astigmatism, field curvature, or distortion. A single aspheric surface may completely correct a single aberration at a single field point. However, complete correction at one field point normally requires aberration at another. An aspheric surface at the lens stop works best over a small angular field, whereas an aspheric surface at the field stop works best at a small NA.

Chapter 5
Basic Physical Optics

5.1 Importance of Physical Optics

Physical optics defines the transport of light by the physics of wave propagation. Maxwell's equations and Fourier transformations are applied toward accurate descriptions of point spread and defocus. Comprehension of the ensuing topics is essential for an effective design of an optical instrument. Development of comprehension is a product of exploration and application of these principles.

Refraction, reflection, and scatter are all based on the wave nature of light. The Airy pattern is dependent on the diffraction of a circular aperture, and the depth of focus is dependent on physical optics.

Most sections of this chapter have a more detailed version in Chapter 19, which focuses on advanced concepts. At least one member of the optical design team should comprehend the advanced concepts.

5.2 Wave Equation

The wave equation in Chapter 19 [Eq. (19.1)] may be reduced to a simple expression for the spatial frequency k and the temporal frequency ω:

$$k^2 = \varepsilon\mu\omega^2 , \tag{5.1}$$

where ε is electric permittivity and μ is magnetic permeability. The electric permittivity is positive for a dielectric and negative for a conductor.

The spatial frequency in radians per distance is

$$k = \frac{2\pi}{\lambda}, \tag{5.2}$$

where λ is the spatial period. The temporal frequency ω in radians per time is expressed as

$$\omega = \frac{2\pi}{T}, \tag{5.3}$$

where T is the temporal period.

The velocity of the wave is

$$\frac{\omega}{k} = \frac{1}{\sqrt{\varepsilon\mu}} = nc, \tag{5.4}$$

where n is the refractive index, and c is the speed of light.

The original language of electromagnetism employed the words "conductor" and "dielectric," as in, "a conductor displays electric current at low frequencies, whereas a dielectric does not." By persistence of this nomenclature, a glass is still considered a dielectric even although it displays significant bound electron current. The bound electron current of a glass defines the refractive index and dispersion of a glass.

Chapter 19 provides more information on the wave equation, electron current, and refractive index.

5.3 Refractive Index

The refractive index n is derived from

$$n^2 = \varepsilon\mu c^2 . \tag{5.5}$$

The spatial frequency within a refractive medium can be expressed as

$$k = n\frac{\omega}{c} = n\frac{2\pi}{\lambda}, \tag{5.6}$$

where λ is the wavelength in vacuum.

The complex refractive index N is expressed as

$$N = n + i\kappa, \tag{5.7}$$

where the real part n is the refractive index and the imaginary part κ is the extinction coefficient. The refractive index is largely dependent on bound charge motion, while the extinction coefficient is dependent on free carrier motion.

Application of the complex refractive index to a plane wave yields the electric field as follows:

$$\mathbf{E} = \mathbf{E}_0 \exp\left[i\left(N\mathbf{k}\cdot\mathbf{R} - \omega t \right) \right], \tag{5.8}$$

where the real part indicates a cosine function and the imaginary part indicates an exponential decay. The complex format facilitates differential operations.

The refractive index n defines an oscillation of the electric field as

$$\mathbf{E} = \mathbf{E}_0 \cos\left(\frac{2\pi}{\lambda} n z - \omega t\right). \tag{5.9}$$

The extinction coefficient κ defines a decay of the electric field as

$$\mathbf{E} = \mathbf{E}_0 \exp\left(\frac{-2\pi}{\lambda} \kappa z\right). \tag{5.10}$$

The extinction coefficient of a glass is normally zero, which implies an infinite depth of the electric field. Conversely, the extinction coefficient of a metal is large, which implies a shallow depth.

The motions of free and bound carriers are quite different. Free electrons move quickly toward cancellation of an external electric field. Consequently, the electric field of a metal decays exponentially toward zero within a fraction of the spatial wavelength. Bound electrons also move toward cancellation of the external field; however, a restoring force of the bond greatly limits cancellation of the external field. Consequently, the electric field of glass returns to zero faster than air, but not within a spatial wavelength. This effect of partial cancellation is expressed by a shortened wavelength.

Chapter 19 describes mathematical models for the refractive index of free and bound electrons. A Drude model applies to free carriers through two parameters: plasma frequency and lifetime. A Lorentz model applies to bound carriers through three parameters: plasma frequency, lifetime, and resonant frequency. In both models, a lifetime indicates the average duration of oscillation before absorption. A metal is described by free carriers, and a glass is described by bound carriers.

Figure 5.1 displays a Drude model for the free electrons of silver. The plasma frequency occurs at 150 nm, and the lifetime is 10 plasma cycles. The refractive index is nearly zero throughout the visible regime, while the extinction coefficient is > 3 from 400 nm and beyond. Most of the reflection is due to free carrier motion without scatter. Consequently, the absorption is nearly zero and the reflectance is nearly 100% [Eq. (5.6)].

Figure 5.2 displays a Drude model for the free electrons of nickel. The plasma frequency occurs at 120 nm, and the lifetime is five plasma cycles. The refractive index grows throughout the visible regime, and the extinction coefficient is > 3 from 400 nm and beyond. The short lifetime of five plasma cycles indicates significant absorption. Consequently, the absorption is large and the reflectance is near 70% [Eq. (5.6)]. Nickel appears darker than silver due to electron scattering, even though the higher plasma frequency of nickel indicates a higher electron density. Silver has superior electron mobility due to less frequent scattering, as is indicated by a longer lifetime. Electron mobility has a profound effect on reflectance.

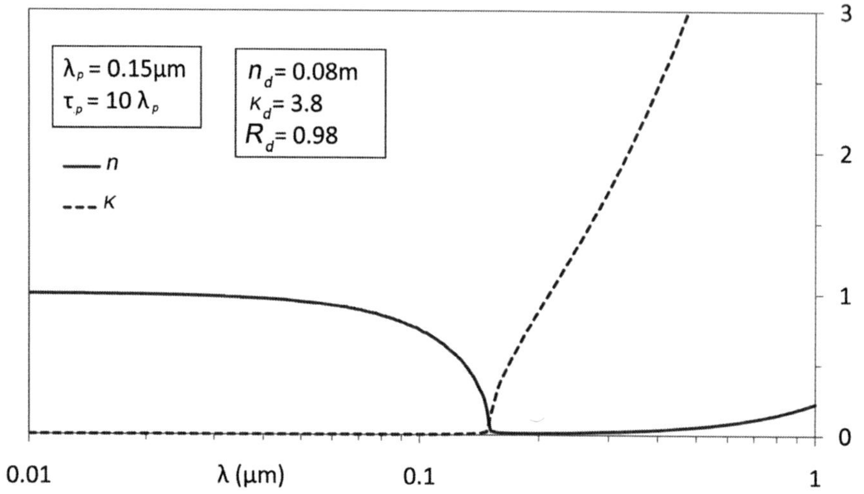

Figure 5.1 Drude model of silver. Refractive index is n, extinction coefficient is κ, and reflectance is R at the d line (589 nm).

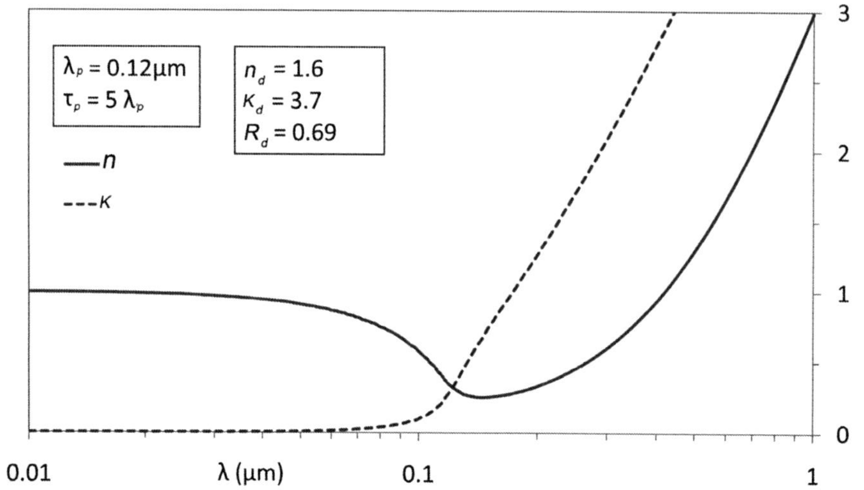

Figure 5.2 Drude model of nickel.

Figures 5.3 and 5.4 display the Lorentz spectra of silica. Both Lorentz models employ a bound resonance at 100 nm. The refractive index resembles the empirical data for silica,[4] while the extinction coefficient varies greatly. The lifetimes are varied for the purpose of exercise of the Lorentz model. In Fig. 5.3, a lifetime at 800 bound cycles creates a reasonable estimate of the refractive index; it grows more rapidly with a shorter wavelength, which defines normal dispersion. In Fig. 5.4, the lifetime is much shorter at 100 bound cycles. The refractive index does not grow more rapidly at shorter wavelengths, which defines anomalous dispersion. Actually, in the example of Fig. 5.4, the refractive index profile is flat from 0.2–1.0 μm, while the extinction profile is broadened.

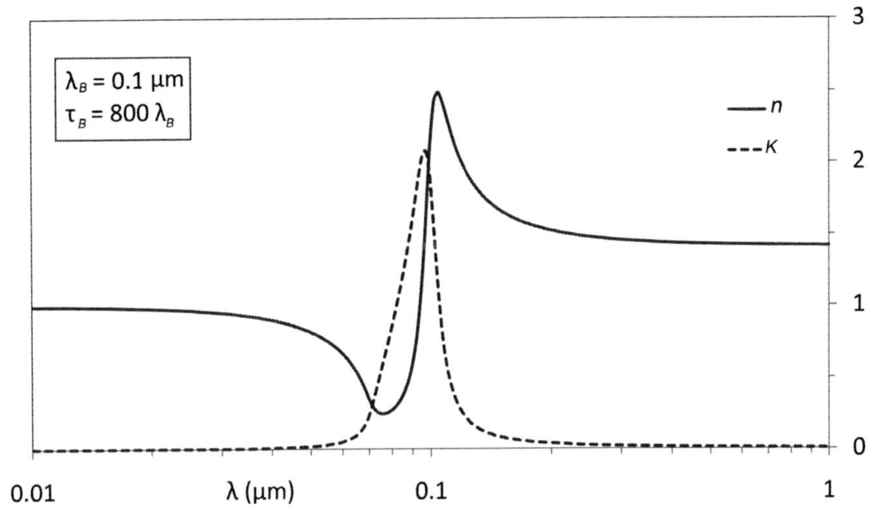

Figure 5.3 Lorentz model of silica.

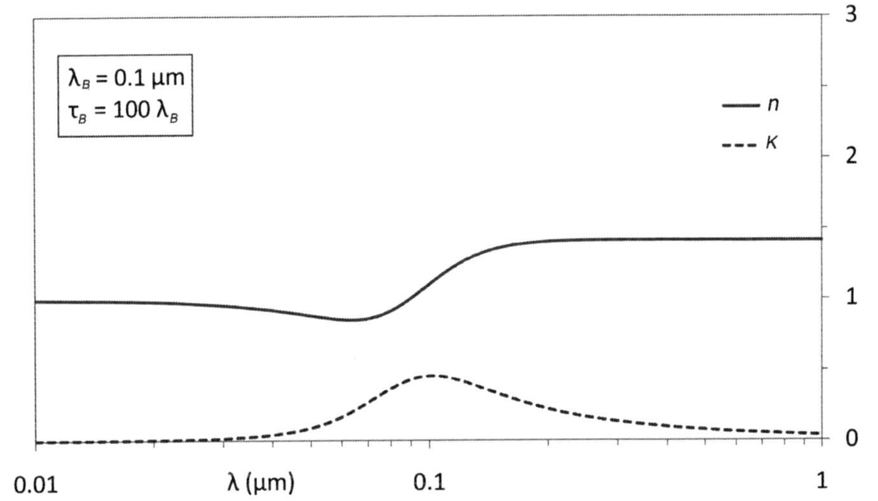

Figure 5.4 Lorentz model of silica with a shortened lifetime. Absorption reduces the magnitude of dipole current and dispersion. A shorter lifetime indicates low dispersion of short flint.

Figure 5.5 displays the transmission of a Lorentz model of silica at different lifetimes, as derived from Eq. (5.9). The internal transmittance is calculated from Eq. (5.21). The lifetime at 800 cycles creates a narrow absorption band, which is largely confined to the ultraviolet regime ($<$ 400 nm). The length of the spectrum is defined by the separation of two absorption bands: a first at 0.1 μm, and a second at 9 μm. The lifetime at 800 cycles defines a long spectrum in the blue, which extends far into the ultraviolet. The lifetime at 100 cycles creates a broad absorption band, which extends into the blue regime (440–490 nm). This defines a short spectrum in the blue, which is a feature of short flint glass. Chapters 17 and 18 provide more information on glass types and refractive indices.

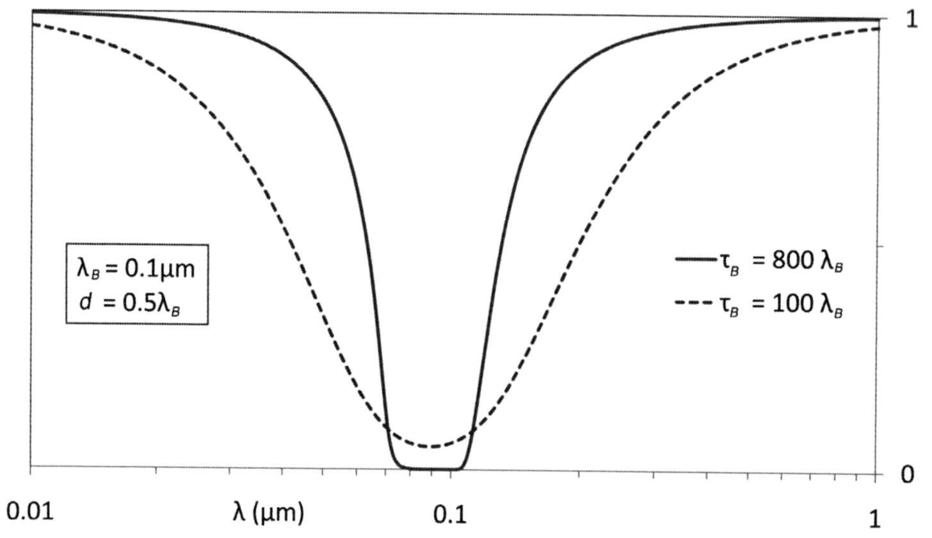

Figure 5.5 Transmittance of a Lorentz model of silica. Absorption increases the width of the absorption band. A shorter lifetime creates absorption of blue wavelengths. A short lifetime displays a short spectrum in the blue.

Aluminum represents a combination of Drude and Lorentz spectra. Its refractive index and extinction coefficient are 1.15 and 7.15 at the d line, respectively. Aluminum behaves like silver with a bound oscillator at 800 nm. Consequently, there is small dip in reflectance at 800 nm. The reflectance of aluminum is near 92% at the d line.

5.4 Dispersion

During normal dispersion, the refractive index grows more rapidly with a shorter wavelength. The Abbe number provides a metric for dispersion:

$$\nu_d = \frac{n_d - 1}{n_F - n_C}, \tag{5.11}$$

where n_d is the d line of helium at 587.6 nm, n_F is the F line of hydrogen at 486.1 nm, and n_C is the C line of hydrogen at 656.3 nm.

The partial dispersion describes the change in refractive index as a fraction of the change between the F and C lines. For example, the partial dispersion of the g and F lines is defined as

$$P_{gF} = \frac{n_g - n_F}{n_F - n_C}. \tag{5.12}$$

The g and F lines are both active in the blue region of color vision. Other spectral lines are listed later in Table 17.6. The SCHOTT glass company defined two glass types as normal in partial dispersion: a crown K7 and a flint F2.

There are two methods for decreasing dispersion ν_d: increasing resonant frequency or decreasing resonant lifetime. A higher resonant frequency places the resonant frequency deeper into the UV; this creates a flat index profile in the visible regime. It also defines a long-spectrum crown, with a large distance between resonant wavelengths. On the other hand, a shorter resonant lifetime represents less dipole current; this also flattens the refractive index profile. A shorter resonant lifetime also indicates absorption in the blue wavelength regime; this defines a short-spectrum flint. Additional information on materials can be found in Chapters 17 through 19.

5.5 Refraction and Reflection

During refraction, the spatial frequencies along the surface must be the same for both materials. Consequently, only the normal component of the propagation vector **k** changes as expressed in Fig. 5.6. The condition of tangential spatial frequencies yields a form of Snell's law as follows:

$$k_1 \sin\theta_1 = k_2 \sin\theta_2 . \quad (5.13)$$

During reflection, the tangential components must remain constant while the magnitude of the propagation vector **k** remains constant. Consequently, the reflection angle must equal the incident angle.

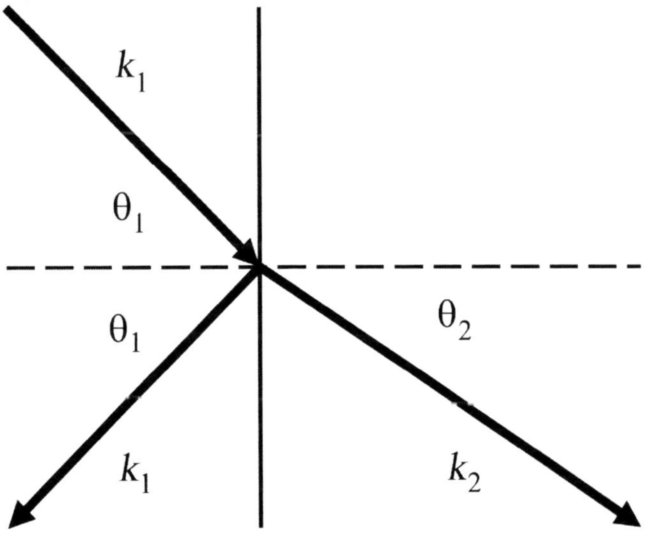

Figure 5.6 Propagation vectors of refraction and reflection.

The Fresnel reflection and refraction coefficients are based on the continuity conditions along the surface. There are two polarizations. A senkrecht polarization specifies an upright electric field to the plane of incidence. A parallel polarization specifies a parallel electric field to the plane of incidence. The reflection coefficients for the electric field of a wave at normal incidence are

$$r_{SN} = -r_{PN} = \frac{N_1 - N_2}{N_1 + N_2}, \tag{5.14}$$

where r_{SN} is the reflection coefficient for senkrecht polarization at normal incidence, and r_{PN} is the reflection coefficient at parallel polarization at normal incidence. A shown in Fig. 5.7, the reflection coefficient flips polarity when the order of materials is reversed. Consequently, a film with a thickness much less than a wavelength displays zero reflection.

The reflectance specifies the reflection of irradiance, which is proportional to the squared magnitude of the electric field. Consequently, the reflectance at normal incidence is

$$R = |r|^2 = \left|\frac{N_1 - N_2}{N_1 + N_2}\right|^2. \tag{5.15}$$

The reflectance at normal is a fair approximation for the sum of both polarizations within the Brewster angle:

$$\tan\theta_B = \frac{N_2}{N_1}. \tag{5.16}$$

The reflectance of a typical glass-to-air interface is 4%, where the refractive index for the glass is 1.5.

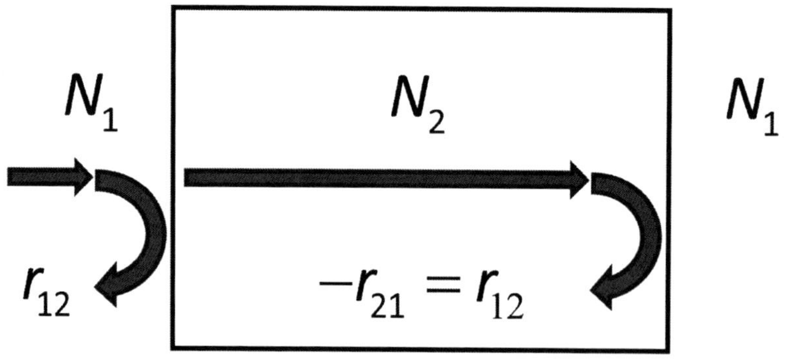

Figure 5.7 Reflection coefficients at normal incidence.

5.6 Emission

Derivation of the irradiance of a dipole emission is

$$I_D = \left(\frac{|\ddot{\mathbf{p}}|^2}{32\,\pi^2 \varepsilon_0 c^3} \right) \left(\frac{\sin^2\theta}{r^2} \right), \tag{5.17}$$

where $\ddot{\mathbf{p}}$ is the temporal curvature of the dipole moment. This pattern resembles a toroid with an inner diameter of zero. The irradiance is zero along the axis of the toroid.

The total radiant flux of a dipole is

$$P = \frac{|\ddot{\mathbf{p}}|^2}{12\pi\varepsilon_0 c^2}. \tag{5.18}$$

If dipole rotation creates an isotropic emission, then the radiant power is evenly distributed over 4π steradians. The irradiance of a spherical emission is

$$I_S = \frac{|\ddot{\mathbf{p}}|^2}{48\,\pi^2 \varepsilon_0 c^3} \frac{1}{r^2}. \tag{5.19}$$

The spherical irradiance is two-thirds the maximum for that of dipole emission.

A dipole can also absorb photon energy where electric-field polarization might be critical. A constrained molecule can display anisotropic absorption with polarization of the excitation field.

5.7 Absorption

Attenuation of the electric field may be expressed through the extinction coefficient as

$$\mathbf{E} = \mathbf{E}_0 \exp\left(-\frac{2\pi}{\lambda} \kappa z \right) \cos\left(\frac{2\pi}{\lambda} nz - \omega t \right). \tag{5.20}$$

The resulting irradiance profile is

$$I = \frac{1}{2} c\varepsilon_0 |\mathbf{E}_0|^2 \exp(-\alpha z), \tag{5.21}$$

where c is the speed of light, ε_0 is the electric permittivity of free space, and α is the absorption coefficient:

$$\alpha = 2\frac{2\pi}{\lambda}\kappa = \frac{4\pi}{\lambda}\kappa. \tag{5.22}$$

The absorption coefficient applies to the electric irradiance, whereas the extinction coefficient applies to the electric field.

5.8 Evanescent Field

The spatial and temporal curvatures display a polarity. The curvature of a sinusoid is negative, which defines an inward curvature toward zero. The curvature of an exponential is positive, which defines an outward curvature from zero. An inward curvature defines an oscillation. An outward curvature defines growth or decay. An evanescent field vanishes over a short distance and is mathematically expressed as positive spatial curvature.

Figure 5.8 displays the crests of the internal and external waves during total internal reflection. There is an internal wave and an external wave. The internal wave obeys

$$k_x^2 + k_y^2 + k_z^2 = k_I^2, \tag{5.23}$$

where k_I is the ordinary spatial frequency of the internal medium. The external wave obeys

$$k_x^2 + k_y^2 - \alpha_z^2 = k_E^2, \tag{5.24}$$

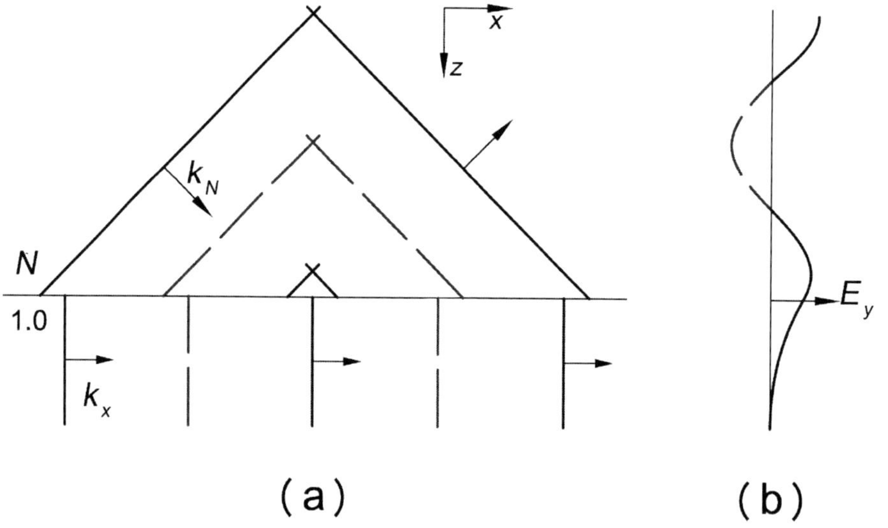

Figure 5.8 Wave crests of total internal reflection: (a) top view of the *xz* plane, and (b) side view of the *xy* plane at the midsection of the top view. A solid line shows a positive peak, and a dashed line shows a negative peak.

where k_E is the ordinary spatial frequency of the external medium. The ordinary spatial frequency is smaller in the external medium. The external wave requires an outward curvature α_z^2, because the sum of the tangential inward curvatures $k_x^2 + k_y^2$ exceeds the inward curvature of the external medium k_E^2. In terms of spatial frequency, the tangential spatial frequency is faster than the external spatial frequency. In terms of wavelength, the tangential wavelength is shorter than the external wavelength. These extraordinary conditions define the exponential decay of an evanescent field.

5.9 Space-Angle Product

An electromagnetic wave does not focus to a point as indicated by a thick-lens model. A ray represents a normal vector to a wavefront. Consequently, an electromagnetic wavefront converges to a point spread. A Gaussian laser beam provides a simple expression for this convergence through the space-angle product:[1]

$$A\Omega = \lambda^2, \tag{5.25}$$

where A is the area of point spread, Ω is the solid angle of convergence, and λ is the spatial wavelength. A Gaussian version of the space-angle product is

$$d_G(2\mathrm{NA}_G) = \frac{\pi}{4}\lambda = 1.27\lambda, \tag{5.26}$$

where d_G is the diameter of the Gaussian beam and NA_G is the Gaussian NA.

The Gaussian space-angle product is based on a fundamental principle of Fourier summation:

$$\sigma_x \sigma_k \geq \frac{1}{2}, \tag{5.27}$$

where σ_x is the standard deviation of spatial position, and σ_k is the standard deviation of spatial frequency.

An extraordinary spatial frequency is defined as

$$k > N\frac{\omega}{c}. \tag{5.28}$$

The existence of an extraordinary spatial frequency requires an evanescent field. Consequently, the NA may exceed a hemisphere in the presence of an evanescent field:

$$\mathrm{NA}_M > n \sin\left(\frac{\pi}{2}\right). \tag{5.29}$$

5.10 Coherence

The axial version of the space-angle product is

$$\Delta z\, \Delta\lambda = \frac{4}{\pi}\lambda_P^2, \tag{5.30}$$

where $\Delta\lambda$ is the wavelength range, and λ_P is the peak wavelength. The coherence length within a refractive medium is

$$\Lambda_C \approx \frac{\lambda_P^2}{n\Delta\lambda}. \tag{5.31}$$

Chapter 19 provides additional information on coherence.

5.11 Airy Pattern

An Airy pattern is created by diffraction of waves at a circular aperture. A plot of an Airy pattern is displayed in Fig. 5.9 along with a Gaussian estimate.

The Airy pattern is frequently defined for an arbitrary aperture and distance.[5] Conversion to marginal NA yields the subsequent relations for optical design. The electric field of an Airy pattern is

$$\mathbf{E}_0 \frac{2J_1(\rho\, k\,\mathrm{NA}_M)}{\rho\, k\,\mathrm{NA}_M}, \tag{5.32}$$

where ρ is the radius about the optical axis, k is the spatial frequency, NA_M is the marginal NA, $\mathbf{E}_0$ is the electric field at the origin, and J_1 is a Bessel function of the first kind of the first order. A Bessel function is a solution to a second-order differential equation with cylindrical boundary conditions.[6] The irradiance of an Airy pattern is expressed as

$$I_{Airy}(\rho) = I_0 \left| \frac{2J_1(\rho\, k\,\mathrm{NA}_M)}{\rho\, k\,\mathrm{NA}_M} \right|^2, \tag{5.33}$$

where I_0 is the peak irradiance (power per area). The first zero crossing of the first-order Bessel function is

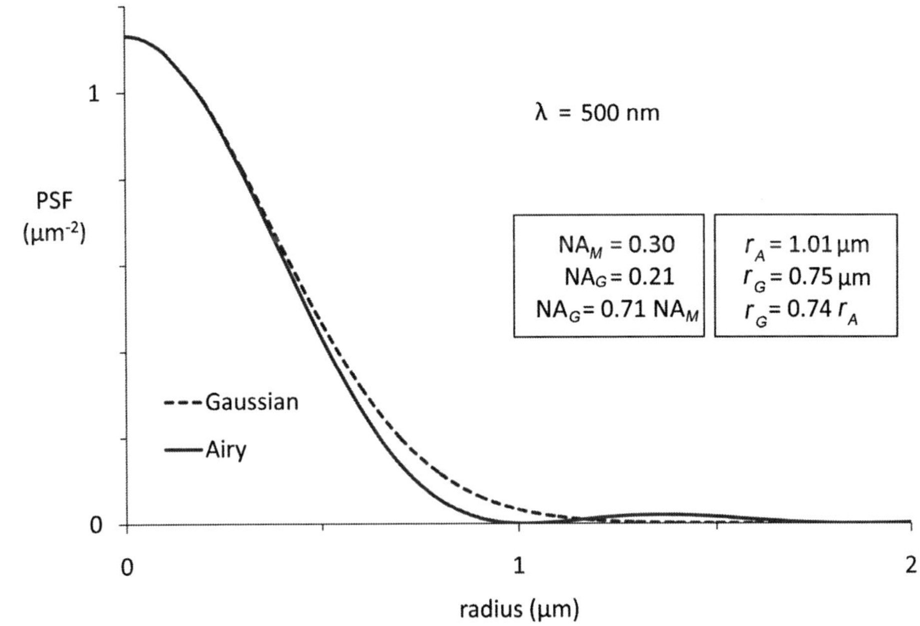

Figure 5.9 Point-spread functions of an Airy pattern and a Gaussian estimate. Standard deviation of a Gaussian estimate toan Airy point spread is < 0.01 in magnitude.

$$\gamma_{J_1} = 3.8317 . \tag{5.34}$$

The diameter of the first dark ring of an Airy pattern is

$$\phi_{Airy} = 1.22 \frac{\lambda}{\mathrm{NA}_M} = \left(\frac{\gamma_{J_1}}{\pi} \right) \left(\frac{\lambda}{\mathrm{NA}_M} \right) . \tag{5.35}$$

The peak irradiance of the Airy pattern is related to the total power P_0 (quanta per time) by

$$I_0 = \frac{\pi \mathrm{NA}_M^2}{\lambda^2} P_0 . \tag{5.36}$$

The space angle product of the Airy disk is derived from Eq. (5.35) as

$$\phi_{Airy} \left(2\,\mathrm{NA}_M \right) = 2.44\lambda = \frac{2\gamma_{J_1}}{\pi} \lambda , \tag{5.37}$$

where NA_M is the marginal NA of the circular aperture responsible for the Airy pattern.

The space-angle product of a linear slit disk is

$$w_{sinc}\left(2\,\mathrm{NA}_M\right)=2.00\lambda\,, \tag{5.38}$$

where w_{sinc} is the width of a sinc function defined as

$$\mathrm{sinc}\left(\frac{x}{a}\right)=\frac{\sin\left(ax\right)}{ax}\,. \tag{5.39}$$

The irradiance of a slit is expressed as

$$I_{slit}\left(x\right)=I_0\,\mathrm{sinc}^2\!\left(xk\mathrm{NA}_M\right), \tag{5.40}$$

where I_0 is the peak irradiance (power per area).

5.12 Gaussian Beam Propagation

A Gaussian profile is the natural mode of a propagating beam of laser light. A plot of a Gaussian beam profile was displayed in Fig. 5.7 as an estimate of an Airy pattern.

The irradiance profile (power per area) is

$$I=I_0\exp\left(-2\frac{r^2}{r_G^2}\right), \tag{5.41}$$

where r_G is the Gaussian radius.

The space-angle product of the Gaussian beam is

$$d_G\left(2\,\mathrm{NA}_G\right)=1.27\lambda=\frac{4}{\pi}\lambda\,, \tag{5.42}$$

where NA_G is the Gaussian NA. All dimensions for space and angle refer to the Gaussian radius r_G.

A Gaussian beam propagates as follows:

$$\phi^2=\phi_G^2+\beta^2z^2\,, \tag{5.43}$$

where ϕ is the beam diameter, ϕ_G is waist diameter, β is the full angle of divergence, and z is the distance of propagation. The waist of the beam is located at $z = 0$. The radius of Gaussian profile r_G defines both the beam diameter and the full angle of divergence.

The radius of the Gaussian wavefront is expressed as

$$R = \frac{1}{z}\frac{\phi^2}{\beta^2}. \tag{5.44}$$

At zero distance, an infinite radius indicates a flat wavefront. At a large distance, the radius R approaches the propagation distance z. The Rayleigh distance is defined as the distance from the beam waist to the point-of-minimum wavefront radius of curvature. As mentioned in Chapter 1, the Rayleigh distance is

$$z_R = \frac{A_0}{\lambda} = \frac{\lambda}{\Omega}, \tag{5.45}$$

where A_0 is the area of the waist. The area of the beam diameter doubles over the Rayleigh distance:

$$A_R = 2A_G. \tag{5.46}$$

The distribution of radiant flux power in a Gaussian profile will be described in Chapter 19.

5.13 Transfer Functions

The MTF of an optical system can be either coherent or incoherent. A laser beam is an example of a coherent system, whereas white light is an example of an incoherent system. A narrowband filter may define a partially coherent system.

The MTF of a coherent system is the amplitude of the aperture function. Thus, a circle creates an MTF in the shape of a cylinder. The cutoff frequency is the spatial frequency of the margin

$$k_M = \frac{\text{NA}_M}{\lambda} = \frac{1.22}{\phi_{Airy}}. \tag{5.47}$$

The MTF of a coherent system applies to the electric field.

The MTF of an incoherent system is the convolution of two cylinders, which resembles a cone. The cutoff frequency of an incoherent system is twice that of a coherent system $2k_M$.

An incoherent MTF can resemble a triangle, where the cutoff frequency is the conical frequency k_C. The conical estimate of an Airy transformation [Eq. (19.34)] defines an effective NA_M from the conical frequency as

$$\text{NA}_M = \frac{k_C \lambda}{1.6}. \tag{5.48}$$

Equation (5.48) is extremely useful in application to MTF data for an off-the-shelf product. Quite often the functional NA is far smaller than the state NA.

Chapter 19 provides more information on the triangle transfer function.

5.14 Gaussian Estimate of Airy Pattern

The peak irradiance of a Gaussian estimate of an Airy pattern is equal to the peak irradiance of an Airy pattern when

$$\sqrt{2}\,\mathrm{NA}_G = \mathrm{NA}_M . \tag{5.49}$$

A Gaussian estimate of an Airy pattern is an effective tool for calculation of important parameters, such as resolution, contrast, and depth of focus. Figure 5.9 displays the point spreads of an Airy pattern and its Gaussian estimate. The standard deviation of a Gaussian estimate from the Airy point spread is < 0.01. Figure 5.10 displays the MTFs of an Airy pattern and its Gaussian estimate. The standard deviation of a Gaussian estimate from the Airy MTF is < 0.05.

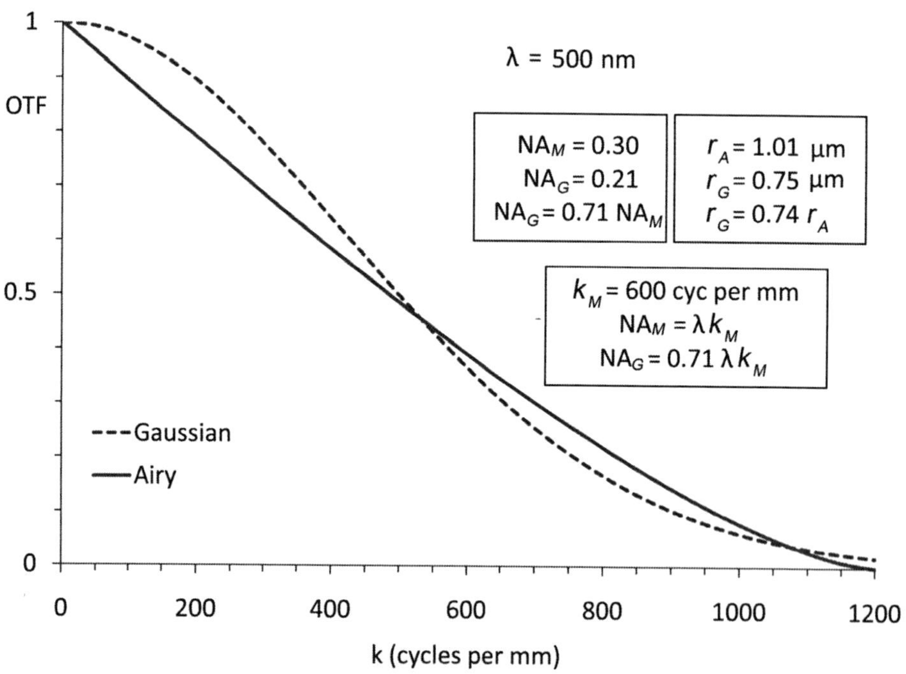

Figure 5.10 Modulation transfer functions of an Airy pattern and a Gaussian estimate. Standard deviation of a Gaussian estimate from Airy MTF is < 0.05 in magnitude.

5.15 Scatter

Surface texture may create a random phase delay during reflection or transmission. The scatter (or scatterance) applies to the irradiance as

$$S = \left\langle \left| s \right|^2 \right\rangle = \phi^2 , \tag{5.50}$$

where ϕ is the rms of the phase delay. The scatterance is frequently called the total integrated scatter (TIS).

At an air-to-glass interface with refractive indices of 1.0 and 1.5, the scatterance of transmission is

$$S_T = 0.25\left(\frac{2\pi\sigma}{\lambda}\right)^2 , \tag{5.51}$$

where σ is the rms of surface texture and λ is the wavelength in air. The scatterance of external reflection is

$$S_{ER} = 4\left(\frac{2\pi\sigma}{\lambda}\right)^2 = 16 S_T . \tag{5.52}$$

The scatterance of internal transmission is

$$S_{IR} = 9\left(\frac{2\pi\sigma}{\lambda}\right)^2 = 36 S_T . \tag{5.53}$$

Obviously, the effect of surface texture is much greater in reflection than in transmission. At 588 nm in wavelength, a 1% scatterance is created by different surface textures: 18 nm for transmission, 5 nm for external reflection, and 3 nm for internal reflection.

Chapter 19 provides more information on surface texture and scatter.

5.16 Interference Filters

An interference filter comprises a dielectric stack of high- and low-index materials. The spectral transmittance or reflectance is transformed by tilt as

$$\lambda' = \lambda \cos\theta . \tag{5.54}$$

At normal incidence, the spatial frequency of reflection is half that of the filter lattice.

Chapter 6
Fluorescence

6.1 Absorption Parameters

The absorption coefficient α applies to irradiance (power per area), while the extinction coefficient κ applies to electric field (force per charge). The transmittance T applies to the irradiance:

$$T = \frac{I_T}{I_0} = \exp(-\alpha z), \tag{6.1}$$

where I_0 is the incident irradiance and I_T is the transmitted irradiance.

The absorption coefficient α is derived from the extinction coefficient κ as

$$\alpha = \frac{4\pi}{\lambda}\kappa . \tag{6.2}$$

The absorption coefficient α is derived from the molar extinction coefficient α_ρ as

$$\alpha = \rho\,\alpha_\rho , \tag{6.3}$$

where ρ is the molar concentration. The units of molar concentration are moles per liter (mol/L). The units of molar extinction coefficient are concentration per distance [moles per liter per centimeter ($\text{mol}\cdot\text{L}^{-1}\text{cm}^{-1}$)]. It is important to correctly define these similar metrics: absorption coefficient α, extinction coefficient κ, and molar extinction coefficient α_ρ.

The absorbance for a small α is defined as

$$A = 1 - T = \alpha z . \tag{6.4}$$

During an actual measurement, reflections of the vessel must be considered:

$$T = (1 - R)\exp(-\alpha z). \tag{6.5}$$

The effect of reflections is eliminated by comparison to a known standard:

$$\frac{T_2}{T_1} = \frac{\exp(-\alpha_2 z)}{\exp(-\alpha_1 z)} = \exp(\alpha_1 - \alpha_2) z. \tag{6.6}$$

$$\left(1 - \frac{T_2}{T_1}\right) = (\alpha_2 - \alpha_1) z. \tag{6.7}$$

6.2 Electron States

The quantum mechanics of absorption and emission are effectively displayed by the electron states of the hydrogen atom.[7] An emission or absorption of a photon requires a change in symmetry as an electron transfers from one state to another. The orbital quantum number j must change by ± 1 for this requirement. The orbital quantum number also describes the angular momentum of electron orbit about the nucleus.

All particles have an intrinsic angular momentum known as spin. The spin quantum number may be $\pm \frac{1}{2}$. The spin of two electrons with identical orbitals must be antiparallel. One electron has "spin up" while the other has "spin down." The antiparallel configuration is called a singlet state because there is only one possible combination of spin components: spin up and spin down. The antiparallel configuration creates a total spin quantum number of 0. During transition to another state, the electron spins remain antiparallel. However, after promotion to another state, an electron spin can be flipped by an external action (EA). The electrons now display a total spin quantum number of 1. The electron spin may now add to the orbital momentum in three different orientations: parallel, perpendicular, and antiparallel. This configuration is a triplet state. The triplet state has a lower energy than the singlet state because the electrons are spinning in the same direction. The triplet state cannot return to the ground singlet state without a spin flip by an EA. Entering a triplet state may permanently stop fluorescence in a process known as "photobleaching."

The angular momentum of an electron state is often described in spectroscopic notation as follows:

$$N^{2s+1}L_j, \tag{6.8}$$

where N is the principal quantum number, s is the total spin quantum number, L indicates the orbital angular momentum by letter (S, P, D, or F), and j is the total angular momentum quantum number $l + s$. The multiplicity of the state is expressed by $2s + 1$. Thus, a singlet state of $s = 0$ has a multiplicity of 1, where j equals the orbital quantum number l, and a triplet state of $s = 1$ has a multiplicity of 3, wherein j equals $l + 1$ (parallel), l (perpendicular) and $l - 1$ (antiparallel).

6.3 Energy Diagrams

A Jablonski energy diagram of a typical fluorophore is displayed in Fig. 6.1. The states are named singlet or triplet with a subscript for excitation level. The ground state is a singlet state S_0. An electron may be promoted to one of two excited singlet states S_1 or S_2. Vibrational states are represented by discrete energy levels of each state. The lifetime of a fluorescent process is typically nanoseconds.

An electron energy is frequently described in units of electron volts (eV). The energy of a photon in electron volts is derived from Planck's law as

$$E = \frac{1240}{\lambda} \text{ eV nm} . \tag{6.9}$$

An EA may convert the excited singlet state S_1 into a similar triplet state T_1 by spin flip of an electron. The parallel spins of the triplet state have lower energy than the antiparallel spins of the singlet state. Consequently, the electron cannot return to S_1 without another EA. The triplet state can return to the singlet ground state S_0 by phosphorescence (P), however, an EA is required for a spin flip during phosphorescence. Consequently, the lifetime of a phosphorescent process is typically seconds.

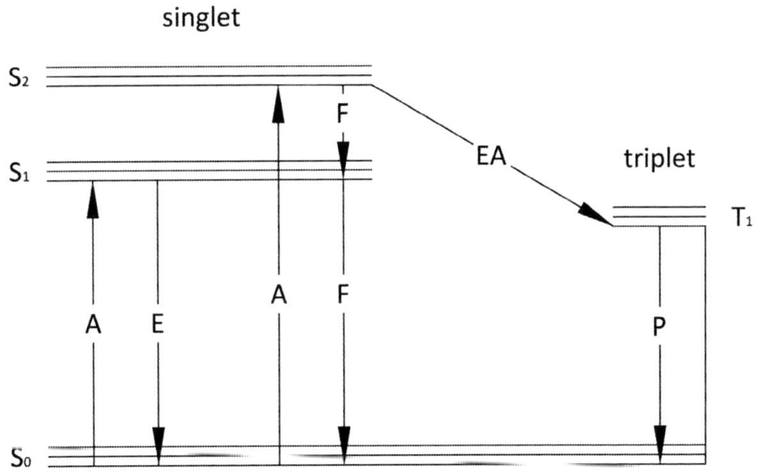

Figure 6.1 Jablonski diagram. S indicates a singlet state with opposing electron spins. T indicates a triplet state with parallel electron spins. Absorption (A) promotes an electron between singlet states. Emission (E) occurs at a similar energy to absorption. Fluorescence (F) indicates an emission at a lower energy than A. An external action (EA) converts a singlet state into a triplet state. Phosphorescence (P) indicates a slow process, which is forbidden without another EA.

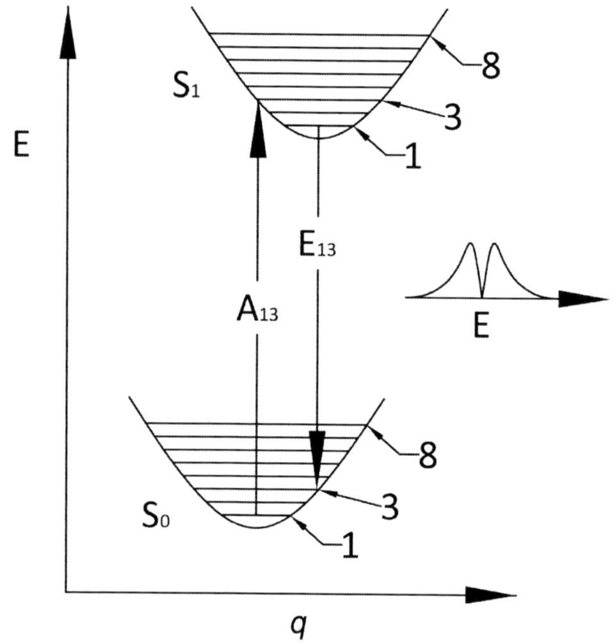

Figure 6.2 Franck-Condon diagram. Electron energy is plotted versus configuration coordinate *q*. S_0 indicates a singlet ground state with antiparallel electron spins. S_1 indicates the first excited singlet state. Numbers 1, 3, and 8 indicate vibration mode. Absorption and emission spectra resemble mirror images.

A triplet state might trigger permanent transformation of the system into a nonfluorescent state. Photobleaching indicates this typical fate of an organic fluorophore. Photobleaching can limit the total emission of a fluorophore to <200 photons.

A Franck-Condon diagram displays the effects of configuration on absorption and emission. In Fig. 6.2, the configuration coordinate q expresses the shape of the molecule in different vibration modes 1 through 8. A Maxwell-Boltzmann distribution describes the population of vibration states as

$$\exp(-\varepsilon_i / kT), \tag{6.10}$$

where ε_i is the energy of a vibration, k is the Boltzmann constant, and T is the temperature. The vibrational state spreads the configuration beyond the electronic state. The spatial overlap of the vibration modes determines the probability of absorption and emission. During absorption, the overlap of configurations grows rapidly with energy, while the opposite occurs during emission. Consequently, the absorption and emission spectra are mirror images of each other.

Figure 6.3 displays the spatial overlaps of configurations in a Franck-Condon diagram. During absorption in Fig. 6.3(a), the overlap grows rapidly with increasing absorption energy until the final vibration state 6 matches the spatial

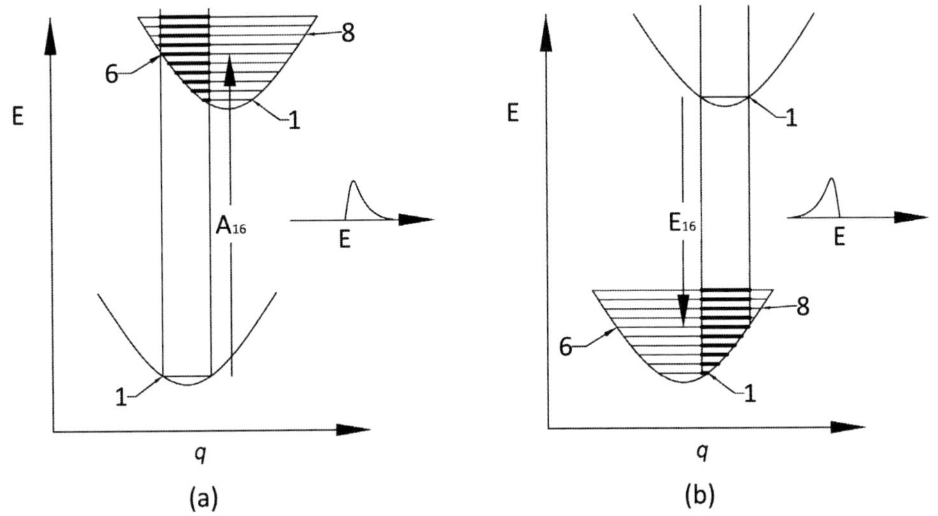

Figure 6.3 Franck-Condon spatial overlap of states. (a) Absorption is dependent on spatial overlap of a single ground state and numerous excited states. Heavy lines indicate overlap of excited states with the lowest ground state. (b) Emission is dependent on spatial overlap of a single excited state and multiple ground states. Heavy lines indicate overlap of ground states with the lowest excited state.

limit of the initial vibration state 1. This indicates an absorption plot that rises rapidly from zero with increasing energy. At greater absorption energy than A_{16}, the overlap decreases by exponential decay. During emission in Fig. 6.3(b), the overlap grows exponentially with increasing emission energy until the final vibration state 6 matches the limit of the initial vibration state 1. At greater emission energy than E_{16}, the overlap of the final state with the initial state decreases rapidly toward zero. If the vibration modes of the ground and excited states are similar in amplitude but different in location, then the emission spectrum is a mirror image of the absorption spectrum. This trend toward mirror image is common in fluorophores.

6.4 Fluorophores

Carbon is the most common element of a fluorophore. A benzene ring C_6H_6 displays a hexagonal structure with strong covalent bonds along the hexagonal plane and weak dipole attractions across the hexagonal plane. Benzene is a building block for organic molecules. Fluorophores are also called chromophores.

Indole is a common organic compound within live organisms. The intrinsic fluorescence of a live organism is frequently due to indole groups. Intrinsic fluorescence is also called autofluorescence. The structure of indole in Fig. 6.4 comprises a hexagonal benzene ring and a pentagonal pyrrole ring. A pyrrole ring contains five elements: four CH groups and one NH group. Neither benzene nor pyrrole is fluorescent;[8] however, their combination as indole is fluorescent.

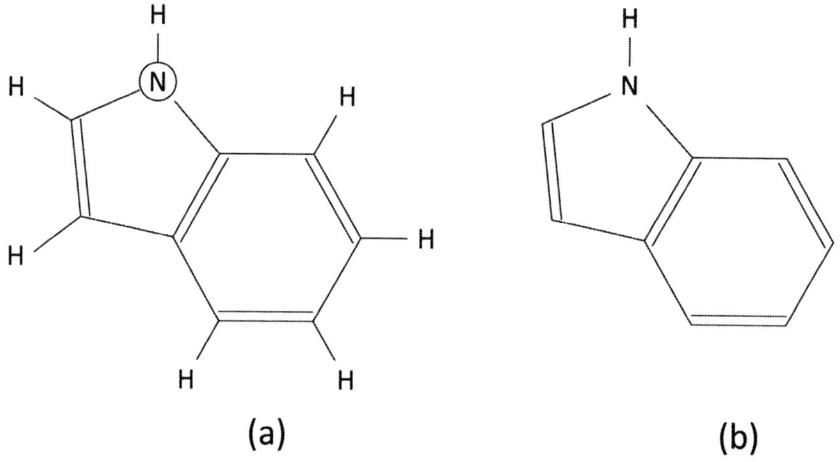

Figure 6.4 Indole structure. Vertices without labels are the carbon sites. Each carbon requires four bonds. (a) HC bonds are displayed. (b) A carbon vertex with three bonds must have a phantom hydrogen (H).

There is a permanent dipole moment from hexagon to pentagon due to the additional proton of the nitrogen element. A polar solvent can align to the molecule with significant effect on the electron states, and also shift or distort emission spectra.

Intrinsic DNA base elements are adenine, thymine, guanine, and cytosine. They display essentially no fluorescence.[9] In Fig. 6.5, adenine is bound to thymine. The weak connections are an attractive force between dipoles. Although this is a "bound state," it is not considered a "bond" due to lack of electron sharing. Adenine absorbs in the UV, but displays insignificant fluorescence. Adenine cannot be used to track its complementary base thymine.

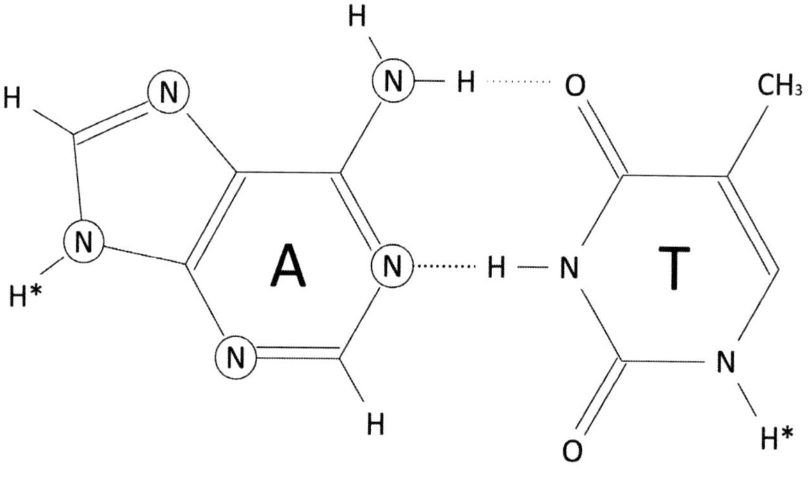

Figure 6.5 Adenine and thymine structure. Adenine and thymine are not fluorescent. Dotted lines indicate weak connections between base pairs of RNA. The H* represent sites that may also be carbon elements of the RNA backbone.

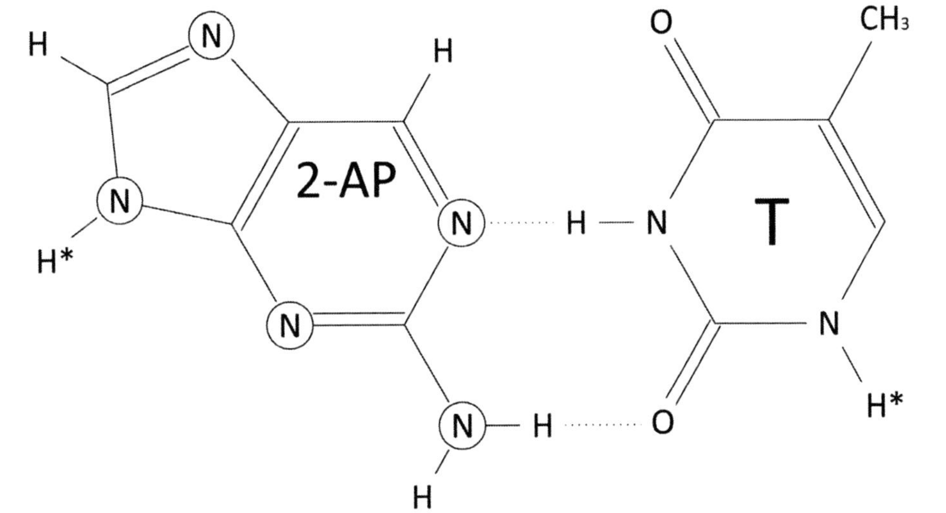

Figure 6.6 2-AP and thymine structure. 2-AP is a fluorescent analog of adenine. Dotted lines indicate weak connections between base pairs of RNA. H* represent sites that may also be carbon elements of the RNA backbone.

An analog of adenine is required as a probe for thymine. As shown in Fig. 6.6, 2-aminopurine (2-AP) is very similar to adenine. An NH_2 group switched location with a CH group. The adenine analog 2-AP absorbs in the UV spectrum and emits fluorescence in the visible spectrum.

Tryptophan is a common protein that contains a fluorescent indole group. The molecular structure of tryptophan is displayed Fig. 6.7, while its absorption and emission spectra are displayed in Fig. 6.8. Tryptophan is a common source of background fluorescence. Fortunately, its emission is near zero at 500 nm and beyond.

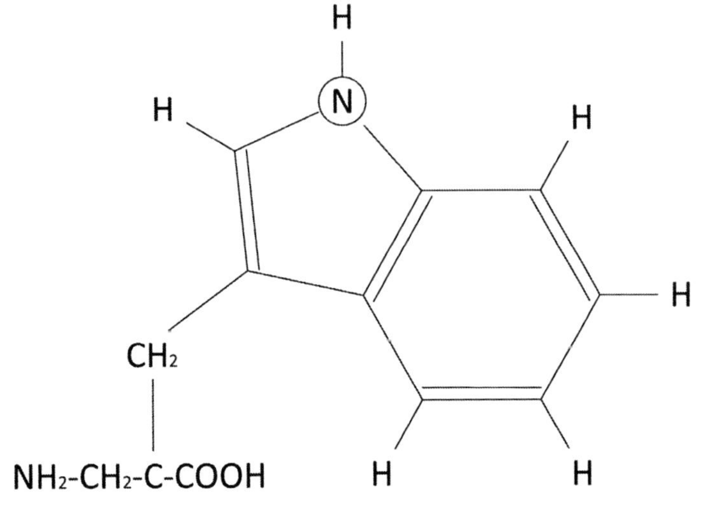

Figure 6.7 Tryptophan structure. An indole group provides fluorescence.

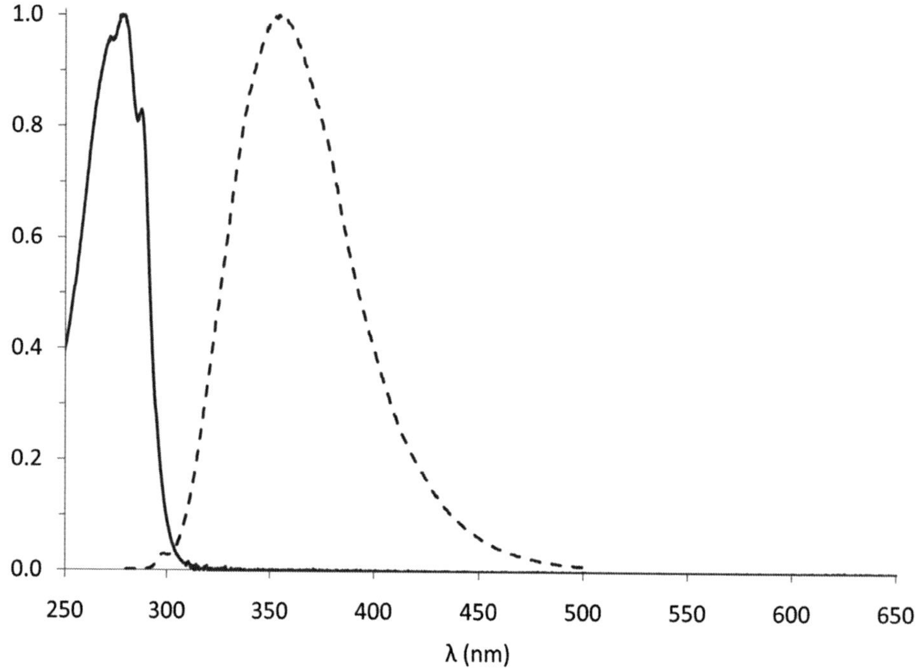

Figure 6.8 Tryptophan spectra. Absorption (solid) and emission (dotted) are nearly mirror images.

Fluorescein is an extrinsic compound to a living organism. The structure of fluorescein is displayed in Fig. 6.9, where the phantom hydrogen atoms are not displayed. The absorption and emission spectra are displayed in Fig. 6.10.

Fluorescein may be "functionalized" by modification of the benzene ring. As shown in Fig. 6.11, a single phantom hydrogen at the bottom benzene ring is replaced with an isothiocyanate group (–N=C=S). The resulting fluorophore is fluorescein isothiocyanate (FITC). The isothiocyanate group is an electrophile that easily bonds to exposed electron pairs. This enables FTIC to bond to other molecules as a fluorescent probe or marker. Other functional groups may be attached, such as antibodies, which may target attached specific proteins.

Green fluorescent protein (GFP) occurs both naturally and synthetically.[9] A natural form of GFP Class 1 exhibits extremely high quantum efficiency: 79% of absorbed photons are converted to emitted photons at 504 nm in peak wavelength. As seen in Fig. 6.12, GFP contains a hexagonal and pentagonal group. Photon absorption ejects a proton from an OH group. This leaves a complex negative anion with an emission in the visible. Synthetic variations of GFP display other colors: yellow 518–529 nm, cyan 490 nm, and blue 440–448 nm. The chromophore of GFP is surrounded by a barrel that is formed by ribbons of protein.

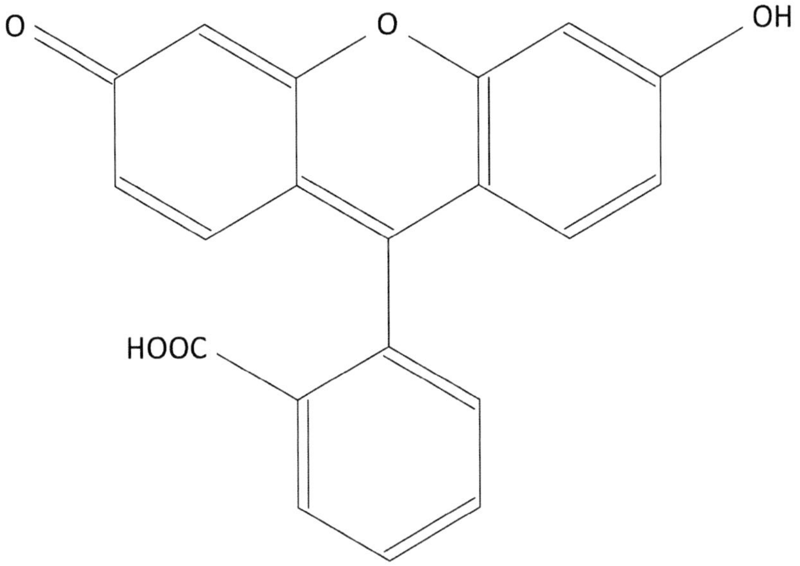

Figure 6.9 Fluorescein structure. A synthetic green fluorescent molecule. A carbon vertex with three bonds must have a phantom H.

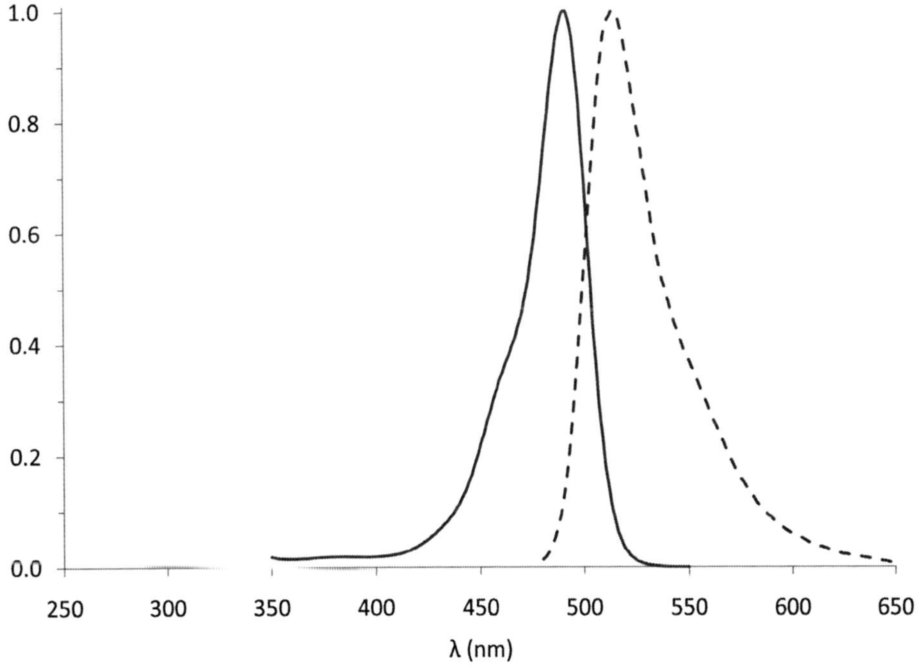

Figure 6.10 Fluorescein spectra. Absorption (solid) and emission (dotted) display significant overlap. Spectra are mirror images.

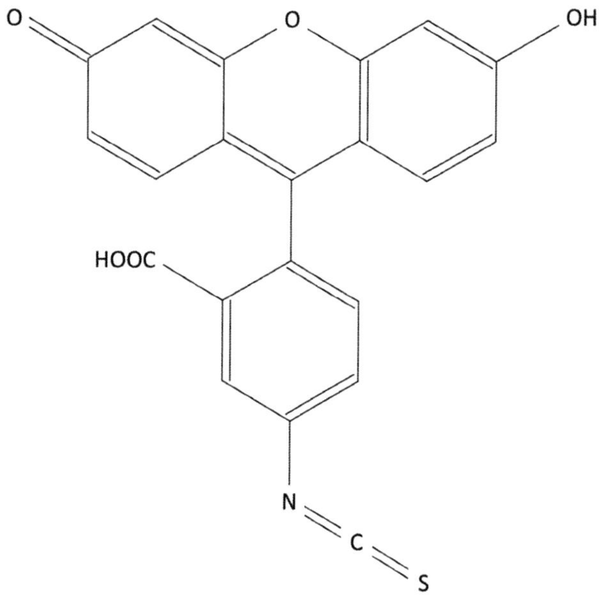

Figure 6.11 Fluorescein isothiocyanate (FTIC) structure. FTIC is a functionalized version of fluorescein. The isothiocyanate group (–N=C=S) bonds easily to exposed electron pairs.

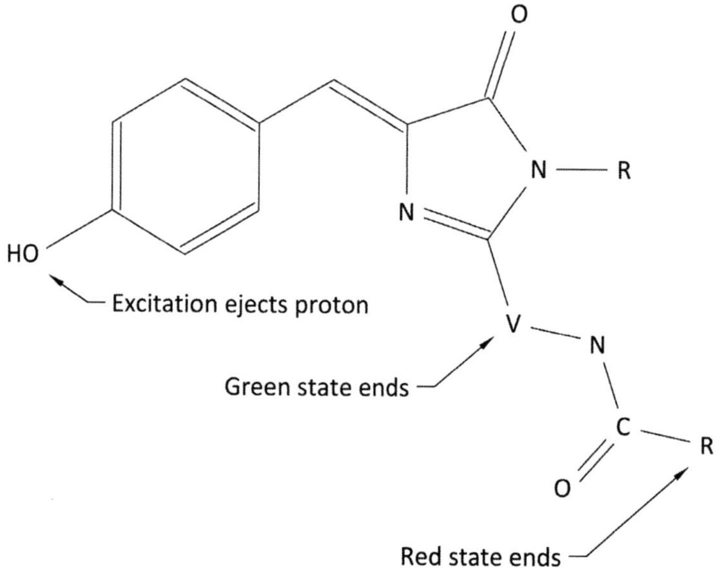

Figure 6.12 Green fluorescent protein (GFP). Excitation of a molecule triggers ejection of a proton. Variable node V determines the spatial extent of an exited state and the wavelength of fluorescence. The excited state of green emission ends at node V. The excited state of red emission ends at ribbon node R. The chromophore sits within a barrel formed by a chain of linear proteins in the form of a ribbon.

Chapter 7
Optical Design Metrics

7.1 CAD Tools

There are numerous genres of computer-aided design (CAD) tools. They offer a variety of metrics in a range from simple illumination calculations to complex wave propagation. It is important to remember that every tool has its limitations. Mastery of these tools requires education and practice.

Various CAD tools are demonstrated in Secs. 7.2–7.9 through application to a doublet with a 40-mm focal length. The doublet functions as a 5X objective in combination with a 200-mm tube lens. The object field height at 1.2 mm translates into a 6.0-mm image height, which accommodates the 11.3-μm diagonal of a 1000 × 1000 array of 8-μm pixels. The projected pixel at the object is 0.8 μm.

The ensuing metrics employ the object side of the lens as the image. The effective object distance (or tube length) is infinite. The entrance pupil is 10 mm. The image NA is 0.0125. The Airy radii are 2.37 μm for the F line (486 nm), 2.86 μm for the d line (588 nm), and 3.20 μm for the C line (656 nm).

Some metrics require sampling of the image. Sampling normally occurs at $2N \times 2N$, where N is an integer. Errors due to insufficient sampling must be considered. The shape of the ray-intercept plot should predict any calculations based on sampling.

7.2 Wavefront Error

The wavefront error is probably the most physically significant metric of a CAD tool. A convergent wavefront should be spherical. A collimated wavefront should be flat. A reflecting telescope with 0.05 NA can easily maintain a wavefront error at $\lambda/10$. Unfortunately, such quality is not easily achieved in a refractor at 0.12 NA or higher.

Figure 7.1 displays the wavefront of the 5X doublet. The optical path difference indicates the optical path length from a perfect sphere around the image. The optical path length is specified in the wavelengths. The abscissa indicates the position within the pupil or lens stop. The wavefront error is independent of the direction of travel through the lens. On reversal of the direction of travel, the optical path difference indicates the optical path length in wavelengths from a perfectly flat wave in collimation. The wavefront error is an effective measure of collimated wavefronts.

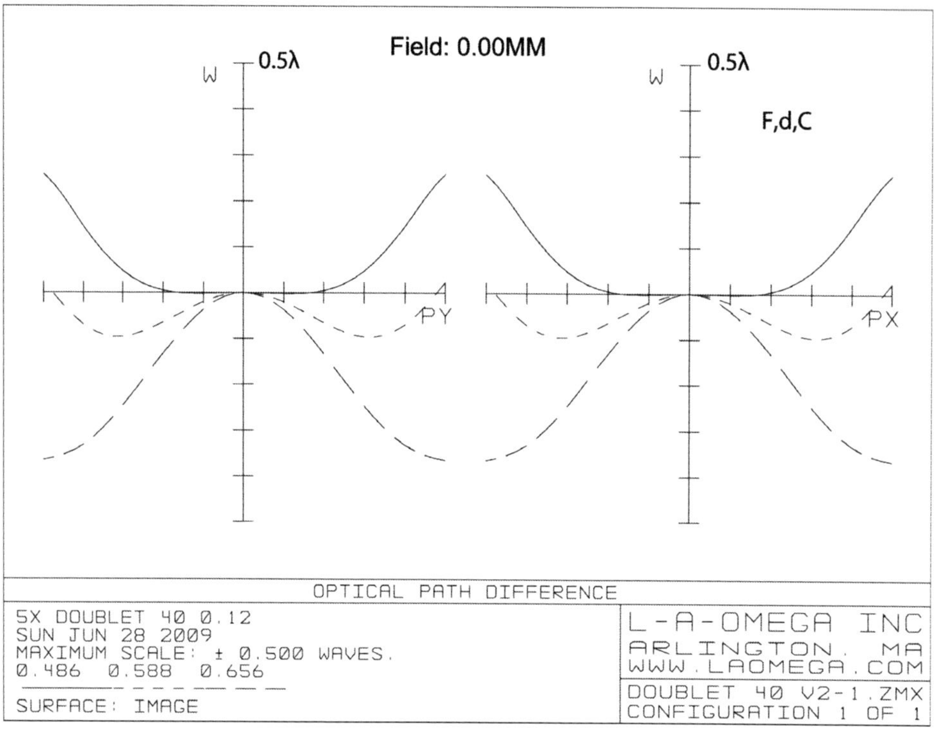

Figure 7.1 Wavefront error of 5X doublet. A solid line indicates the F line (486 nm). A short dash indicates the d line (589 nm). A long dash indicates the C line (656 nm).

The wavefront error of Fig. 7.1 reveals several important features. The d line (588 nm) indicates a wavefront error of 0.1λ, which is considered perfect within the limits of an Airy pattern. The p^4 shape of the F line (486 nm) indicates a fourth-order spherical aberration in wavefront error. The positive wavefront error of the margin indicates a marginal convergence greater than the perfect optimum. The negative wavefront error of the C line (656 nm) indicates axial color, where the convergence is less than the perfect optimum.

A collimating optic, such as a beam expander or a laser, is normally qualified by a wavefront error, which may be calculated and/or measured. A CAD tool may define fabrication and assembly tolerances for an acceptable wavefront error. A wavefront analyzer may provide empirical data as confirmation of wavefront quality. The wavefront error is an essential parameter for application of laser beams.

7.3 Ray-Intercept Plot

The ray-intercept plot (also known as a ray fan) is an essential tool for optical design. It quantifies the geometric point-spread error as a function of ray position within the aperture or lens stop. The geometric point-spread error may be expressed in units of space or angle. The ray intercept is typically specified in microns (μm).

Originally, the terms sagittal and tangential referred to cylindrical coordinates. The sagittal vector represented the radius, as in an archer shooting arrows various directions from the same point. The tangential vector represents the tangent of the circle to the head of the sagittal vector. The tail of the tangential vector is located at the head of the sagittal vector, and the tangent vector is parallel to the circle. In medicine, a sagittal plane splits the body into left and right portions. in fact, a sagittal plane normally splits a system into two symmetrical parts; however, this is not so in the geometric optics of 2009.

The orientation of a sagittal plane of geometric optics is opposite from other conventions. The tangential plane (*yz*) of geometric optics contains the chief ray and the optical axis. The tangential plane may also be called the *y* plane. The sagittal plane (*xz*) of geometric optics contains the chief ray and is normal to the tangential plane. The sagittal plane may also be called the *x* plane. The sagittal plane of geometric optics does not split an optical system into two symmetric parts.

Figure 7.2 displays the ray intercept of the 5X objective. The error is plotted for two orthogonal planes. The object field height is normally specified within the tangential plane *yz*. The tangential error EY is plotted versus the tangential pupil position PY. The sagittal error EX is plotted versus the sagittal pupil position PX.

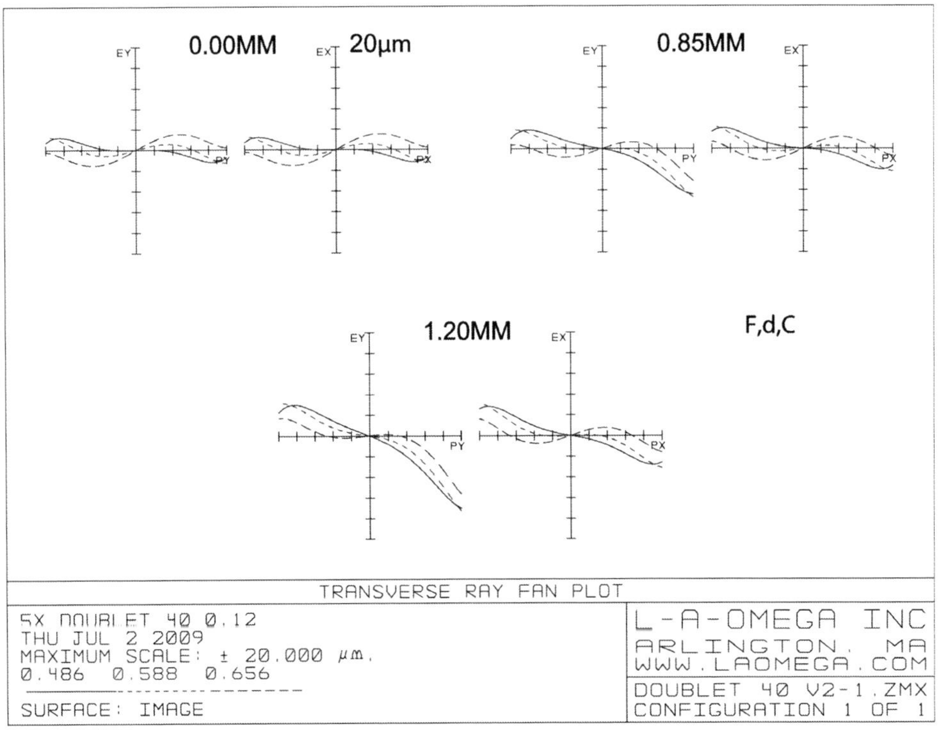

Figure 7.2 Ray-intercept plot of a 5X doublet. The tangential error EY is plotted versus the tangential pupil PY. The sagittal error EX is plotted versus the sagittal pupil PX. A solid line indicates the F line (486 nm). A short dash indicates the d line (588 nm). A long dash indicates the C line (656 nm).

The ray intercept of Fig. 7.2 reveals several important features. The fans are normally defined for three different positions: the on-axis position at zero field height, the margin of the field, and 0.7 times the margin of the field, which represents half the field area.

The on-axis F line (486 nm) indicates a paraxial focus through a flat slope in the central pupil zone. However, the marginal portion of the F line indicates both third- and fifth-order spherical aberration. The on-axis d and F lines both display third-order spherical aberration. The slopes of the F-line plots decrease steadily with field height. This indicates the defocus of Petzval curvature. The dissimilar slopes of F-line plots between EY and EX indicate astigmatism. A small downward curvature in the tangential F-line plots indicates coma. The astigmatism is larger than the coma.

7.4 Spot Diagram

A spot diagram is an effective representation of optical point spread on a large screen with color display. It quickly displays the point spread with reference to a pixel or an Airy disk. A spot diagram is an effective graphic to display to personnel without expertise in optical design. Ordinarily, the spots of three wavelengths are displayed as blue, green, and red; however, spot diagrams are largely absent from this book due to the limitations of monochrome printing.

Figure 7.3 displays a spot diagram of the 5X doublet. A circle displays the Airy diameter at each field point. The on-axis rays are barely contained within the Airy disk. The elliptical shape of the off-axis points indicates astigmatism. A pointed end of an ellipse indicates coma. The bottom header specifies three important parameters: the Airy radius, the rms radius, and the geometric radius. The Airy radius indicates a diffraction-limited resolution, the rms radius indicates an aberration-limited resolution, and the geometric radius indicates the entire spread of the rays.

7.5 Point-Spread Plot

A CAD tool can assign a uniform or Gaussian profile to the pupil. A point-spread plot includes the effects of wave propagation. Planar waveforms are summed to create the point spread. The optical path difference of each surface is incorporated as phase information in the summary. Sampling arrays determine the accuracy of the point-spread plot. The accuracy degrades rapidly with phase delays beyond a wavelength.

Figure 7.4 displays a point spread for the 5X objective. The point spread is monochromatic, and it has large aberration. It is not a perfect Airy pattern. Consequently, the central peak is displayed as less than unity. This central peak specifies the Strehl ratio, which is the ratio of the peak irradiance of point spread to the corresponding peak of the Airy pattern. The Strehl ratio for the off-axis

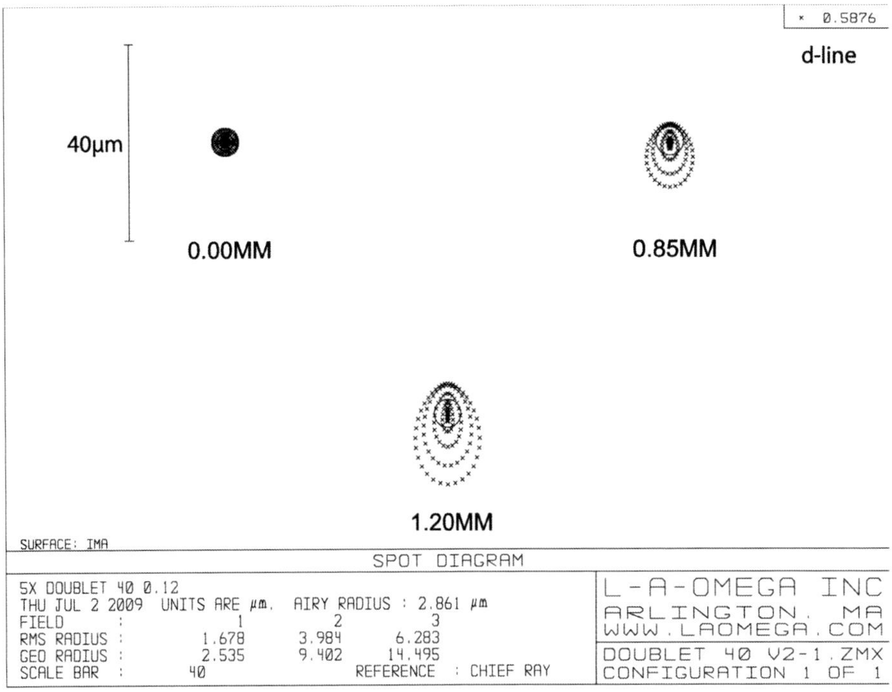

Figure 7.3 Spot diagram of a 5X doublet. Wavelength is at the d line. Ray locations are indicted by ×. The Airy disc is defined by a circle within the field at 1.2 mm.

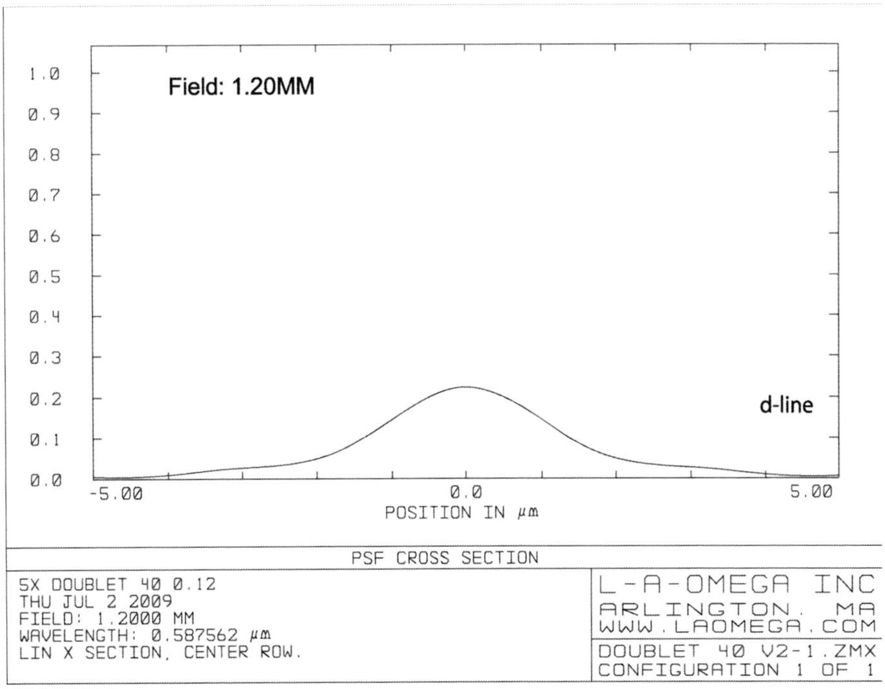

Figure 7.4 Point spread of a 5X doublet that is monochromatic at d line. The field is at 1.2 mm. The Strehl ratio is 21%. Sampling is at 128 × 128.

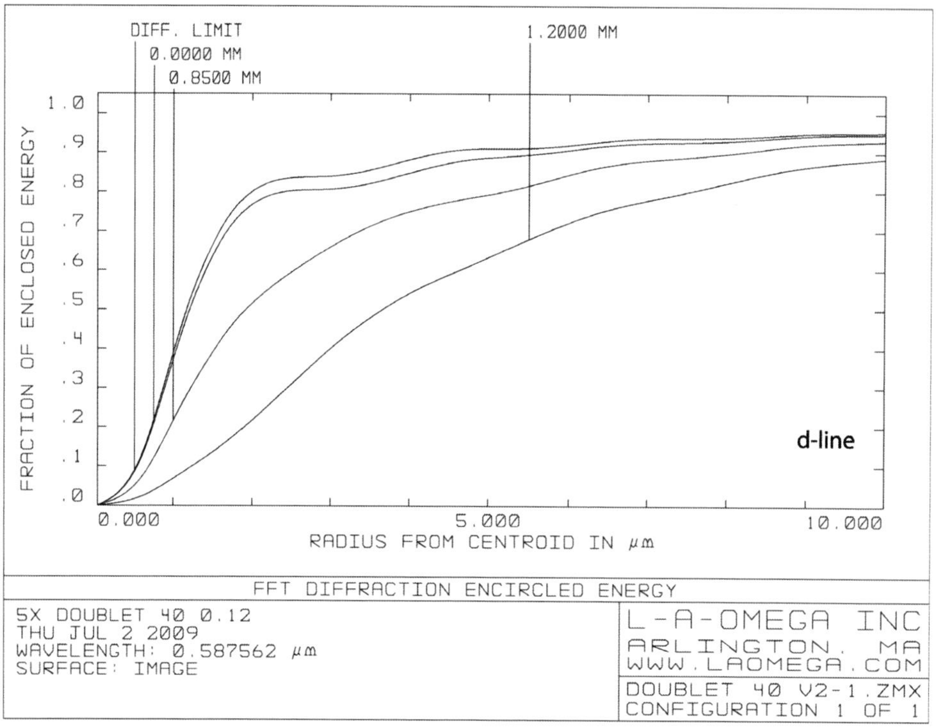

Figure 7.5 Encircled-energy plot of a 5X doublet that is monochromatic at the d line. Sampling is at 128 × 128.

monochromatic point spread is 21%. A point spread is frequently normalized to display the shape. The scale of a point spread must be scrutinized.

7.6 Encircled-Energy Plot

Encircled energy is derived from the point-spread plot. It is an essential metric for low-light applications, where the collection of light should be maximized within a block of four pixels. An encircled-energy plot is far more useful than a point-spread plot.

Figure 7.5 displays an encircled-energy plot for a 5X doublet. The monochromatic encircled energy at 4 μm varies greatly by field: 85% on axis, 75% at 0.85 mm, and 54% at 1.20 mm. The diffraction limit indicates 88% at 4 μm. The flat spot on the diffraction-limit plot line indicates an Airy radius at 2.8 μm.

7.7 Modulation Transfer Function

Figure 7.6 displays a modulation transfer function (MTF) for the 5X doublet. A single field is shown because multiple fields can overlap in a confusing manner. The conical frequency [further discussed in Eq. (19.34)] at 100 cycles/mm indicates a central NA of 0.037 at the d line (588 nm) [Eq. (5.29)]. The paraxial NA is 30% of the central NA.

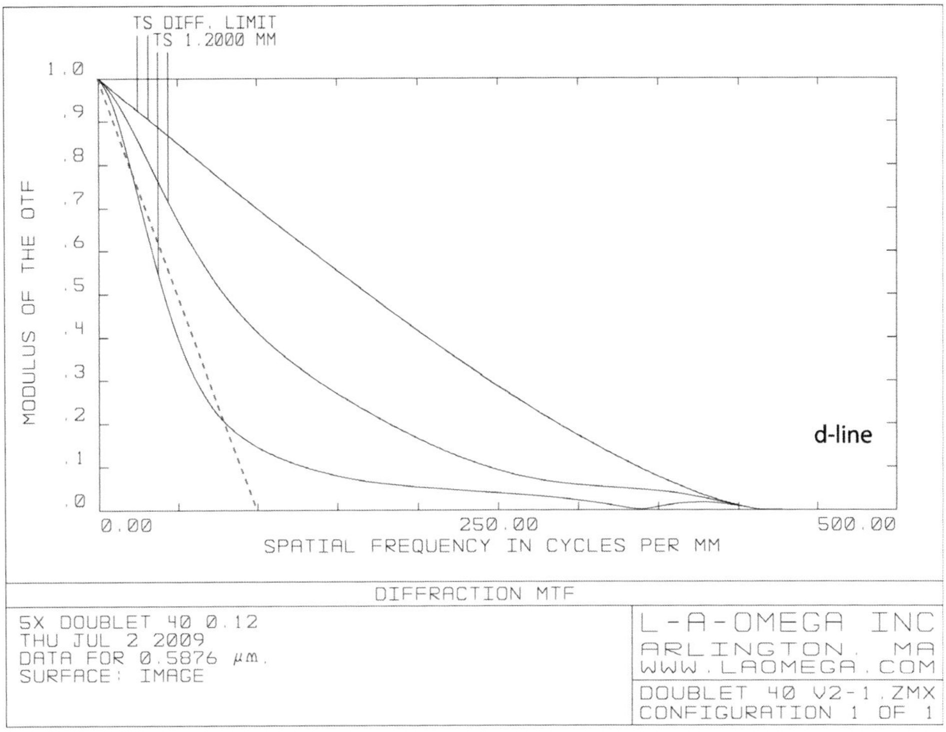

Figure 7.6 MTF of a 5X doublet, where T is the tangential plane and S is the sagittal plane. The modulus is the modulation index. Sampling is at 64 × 64 along the plot scale. A conical approximation (dashed line) indicates a conical frequency at approximately 100 cycles per millimeter.

Some vendors provide MTF data for lenses, while many do not. A conical approximation may be derived from published data. Ronchi rulings may provide empirical measurement at spatial frequencies up to 500 line pairs/mm.

7.8 Edge Spread

An edge spread is an effective measurement for optics with large NAs. Spherical aberration can spread the base of point spread beyond recognition. Observation of a high-contrast edge can reveal important aberrations that are not apparent from a point spread.

Figure 7.7 displays the edge spread of the sagittal edge along the tangential plane for the d line at 1.20 mm. The edge spread indicates 10% at –8.0 µm and 90% at 4.5 µm. The leading edge is much larger than the Airy radius, while the trailing edge is approximately twice the Airy radius.

7.9 Lens Report

A lens report displays several metrics in one graphic. A report can be easily pasted into a design log, such as a Microsoft Word file. Figure 7.8 displays a

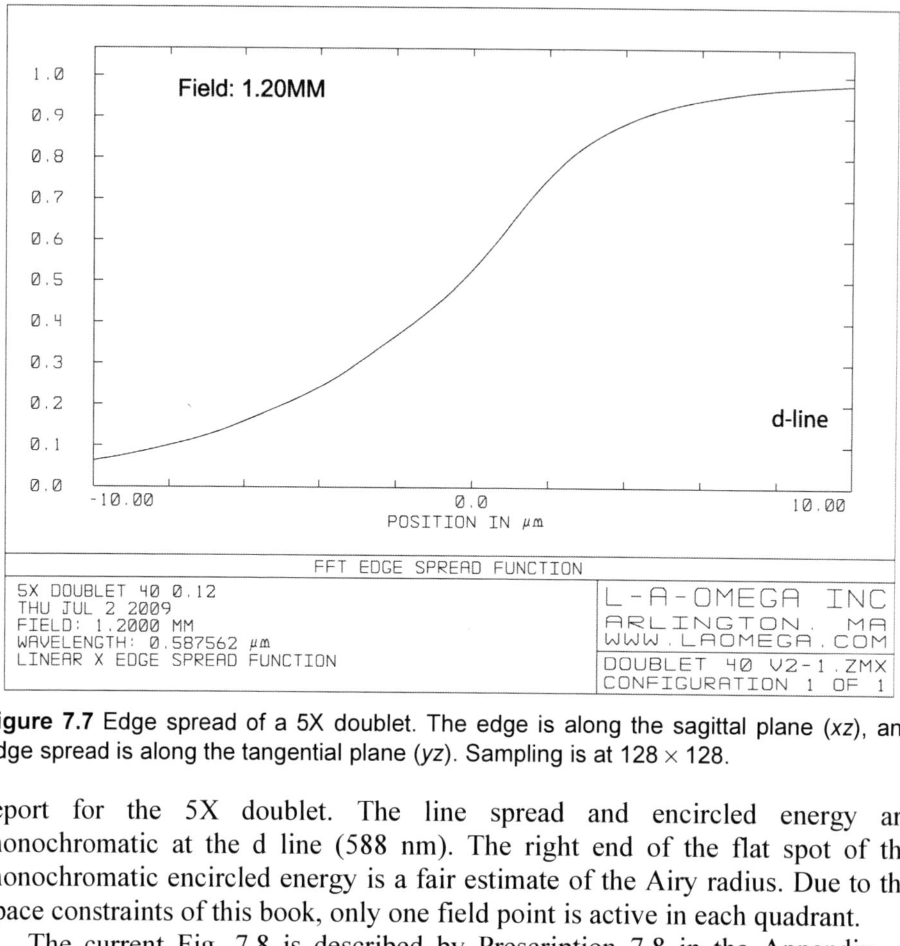

Figure 7.7 Edge spread of a 5X doublet. The edge is along the sagittal plane (*xz*), and edge spread is along the tangential plane (*yz*). Sampling is at 128 × 128.

report for the 5X doublet. The line spread and encircled energy are monochromatic at the d line (588 nm). The right end of the flat spot of the monochromatic encircled energy is a fair estimate of the Airy radius. Due to the space constraints of this book, only one field point is active in each quadrant.

The current Fig. 7.8 is described by Prescription 7.8 in the Appendix. A boxed cell indicates a variable during operation. (A corresponding prescription to every lens report is located in the Appendix. The prescription numbers correlate to the figure numbers.)

7.10 Relative Illumination

A relative illumination plot displays the effects of reduction of NA with field angle. A dependency on $\cos^3\theta$ is derived from the solid angle of collection. The addition of a Lambertian radiance creates a dependency on $\cos^4\theta$. Numerous lens types (such as a double Gauss) reduce NA at the field margin to control astigmatism. Consequently, the effective *f*/# of a double Gauss inflates rapidly with field angle, and *f*/# inflation reduces brightness.

7.11 Surface-Form Error

An interferometer may qualify a surface as round or flat within a fraction of a wavelength. A surface-form error indicates a deviation from a reference surface

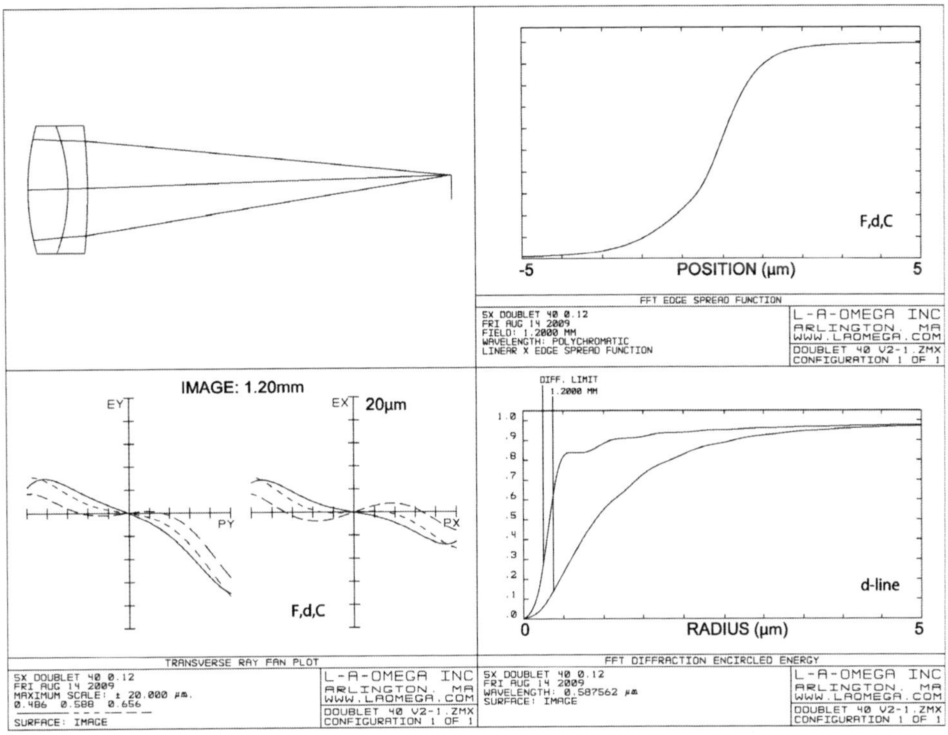

Figure 7.8 Lens report of a 5X doublet. Top left: layout. Top right: line spread. Bottom left: ray intercept. Bottom right: encircled energy. A single point at the field margin (1.2 mm) is employed for each quadrant.

by wavelengths or fringes. Optical elements are fabricated within tolerances for surface-form error at a test wavelength.

The fringes must be converted to optical path lengths to calculate image quality. An optical path length of 0.1λ is considered near perfect in practice. In transmission, the number of test fringes N_F converts to optical path length in cycles as

$$\Lambda_T = \frac{\Delta n N_F}{2}\left(\frac{\lambda_T}{\lambda_A}\right)\text{cyc}, \tag{7.1}$$

where Δn is the change in refractive index, λ_T is the test wavelength, and λ_A is the application wavelength. In reflection, the test fringes convert optical path length as

$$\Lambda_R = n N_F\left(\frac{\lambda_T}{\lambda_A}\right)\text{cyc}, \tag{7.2}$$

where n is the refractive index. Obviously, the effective surface-form error is much greater in reflection than in transmission.

7.12 Manufacturing Standards

The International Standards Organization (ISO) provides standards for the specification of optical systems. The ISO standards are mainly comprised of the old standards of two organizations, the American National Standards Institute (ANSI) and Deutsche Industrie Normen (DIN). The title of the ISO standard is *ISO 10110–Preparation of drawings for optical elements and systems.*

Volume defects of a material of an optical element are specified within three categories: 0, birefringence; 1, bubbles and inclusions; and 2, inhomogeneity and striae. Birefringence indicates the anisotropy of the refractive index. Bubbles and inclusions specify the acceptable limits for voids and extrinsic solids. Inclusions may be crystalline blocks that may gouge the surface during polishing. Inhomogeneity expresses the variation of the refractive index. Striae frequently occur in the shape of filaments or cords.

Surface qualities are specified in several categories: λ, surface coating; 3, surface form; 4, centering; 5, surface imperfections; and 6, laser irradiation damage threshold. The surface coating is typically a thin-film coating. The surface form specifies the deviation of the surface from perfection in fringes. The centering of the surface specifies either a tilt or a lateral displacement of the surface from the optical axis. The surface imperfections specify the acceptable limits of scratches, digs, and chips.

Surface texture defines the fine structure of the material. It may be expressed as an rms of the texture width, which is similar to the depth, or as a power spectrum density, wherein the rms amplitude per spectral range is cited.

Chapter 8
Image Contrast

8.1 Radiometry

Radiometry generally applies to the distribution of quanta over space, angle, and time. Specifically, radiometry applies to energy, whereas photometry applies to luminous energy, measured in Talbots. In the current discussion, radiometry applies to both. The terminology is defined in Table 8.1.

Within the field of optics, flux is generally accepted as flow. However, in physics, flux indicates the strength of a field per area. Sensors for commercial applications are frequently defined by luminance in lux, whereas sensors for scientific applications are normally specified in photons or electrons. Quantum efficiency indicates the conversion of photons to electrons.

8.2 Expression of Contrast

The contrast of an object is defined as

$$C = \frac{\Phi_{OP} - \Phi_{BP}}{\Phi_{BP} + \Phi_{D}}, \tag{8.1}$$

where Φ_{OP} is the flux of the object pixel, Φ_{BP} is the flux of the background pixel, and Φ_D is the flux of the detector. The range of contrast is from -1 to $+\infty$. A negative contrast indicates a dark object within a bright background. A positive contrast indicates an object that is brighter than the background. Quite often both objects are rather dark.

The flux of the detector has numerous components, as described in Chapter 15. The two most common are dark current and read noise. They are often stated as electron currents. The flux of the detector may be expressed as

$$\Phi_D = \frac{\Phi_{DC} + \Phi_{RN}}{\eta_Q}, \tag{8.2}$$

where Φ_{DC} is the mean flux of the dark current in electrons per pixel, Φ_{RN} is the mean flux of the read noise in electrons per pixel, and η_Q is the quantum

Table 8.1 Terms and symbols of radiometry.

Name	Symbol	Description	Units
Quanta	Q	Quanta	photon
		Luminous energy	joule (J)
		Radiant energy	Talbot (T)
Fluence	F	Quanta per area	quanta per area
			joules per area
			Talbot per pixel
Flux	Φ	Quanta per second	quanta per second
		Optical power	flow
		Radiant flux	watts (W)
		Luminous flux	lumens (lm)
Directance	D	Quanta per angle	quanta per steradian
		Radiant intensity	watts per steradian ($W \cdot sr^{-1}$)
		Luminous intensity	lumens per steradian ($lm \cdot sr^{-1}$)
			candela (cd)
Exitance	E	Exiting flow per area	flow per area
			flow per pixel
			watts per area ($W \cdot cm^{-2}$)
			lumens per area ($lm \cdot m^{-2}$)
			lux (lx)
Incidence	I	Incident flow per area	flow per area, flow per pixel
		Irradiance	watts per area ($W \cdot cm^{-2}$)
		Illuminance	lumens per area ($lm \cdot m^{-2}$), lux (lx)
Radiance	L	Flow per area per angle	flow per pixel per hemisphere
		Radiance	watts per area per angle ($W \cdot cm^{-2} sr^{-1}$)
		Luminance	candelas per area ($cd \cdot m^{-2}$)

efficiency of conversion of photons into electrons by the detector. The dark current is normally thermal promotion of carriers into the conduction band. The read noise is created during conversion of electrons into voltage. There are numerous other sources of noise.

Visibility is defined as

$$V = \frac{\Phi_{max} - \Phi_{min}}{\Phi_{max} + \Phi_{min}}. \tag{8.3}$$

Its range is from 0 to 1. It is also the modulation index of the MTF.

Background radiance must be carefully considered. It may originate from the foreground or background of the object, or even from the optical path. The cover strata might emit fluorescence.

Detector noise must also be considered. Dark current, read noise, and dark noise are three common metrics of electronic detectors of light. The object signal may overcome the read noise through integration. However, the object signal cannot overcome the dark current by integration. The dark current and read noise should be extracted from the dark noise of a detector specification. The shot noise of the dark current and the object signal should also be considered.

8.3 Shot Noise

A shot is a fixed quanta, such as a lead sphere, a photon, or an electron. It is not possible to detect a portion of a shot. If the expected mean photon count is 2 photons, then the actual count can be 0, 1, 2, 3, or more. The probability of a count is described as a Poisson distribution:

$$P(n) = \frac{\mu^n \exp(-\mu)}{n!}, \tag{8.4}$$

where μ is the mean and n is the count. The Poisson distribution for a mean dark current of 10 electrons is displayed in Fig. 8.1, where the count ranges from 5 to 15. The Poisson distribution for a mean dark current of 100 electrons is displayed in Fig. 8.2, where the count ranges from 80 to 120.

As the mean count increases, the standard deviation of the shot count approaches the square root of the count. The flux of the shot noise may be expressed in arbitrary units of flux as follows:

$$\Phi_{SN} = \sqrt{q_S \Phi} \quad \text{as } Q > 1000, \tag{8.5}$$

where q_S is the discrete quantum of the shot, Φ is the mean flux, and Q is the expected quantum. The contrast with shot noise grows by the square root of the expected quantum Q:

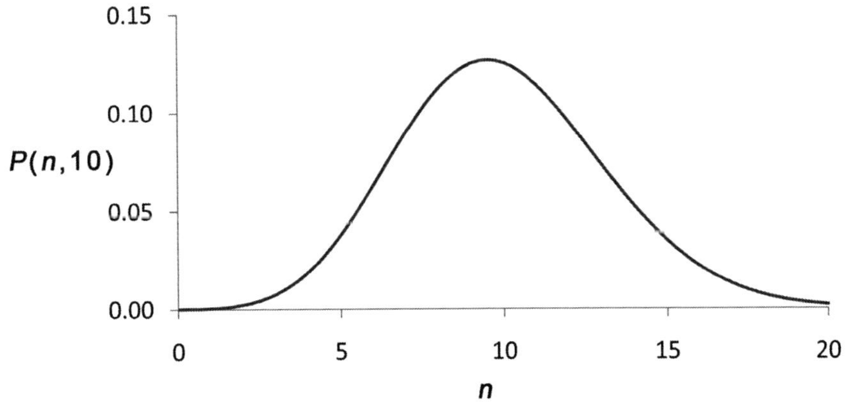

Figure 8.1 Poisson distribution for a mean quanta of 10.

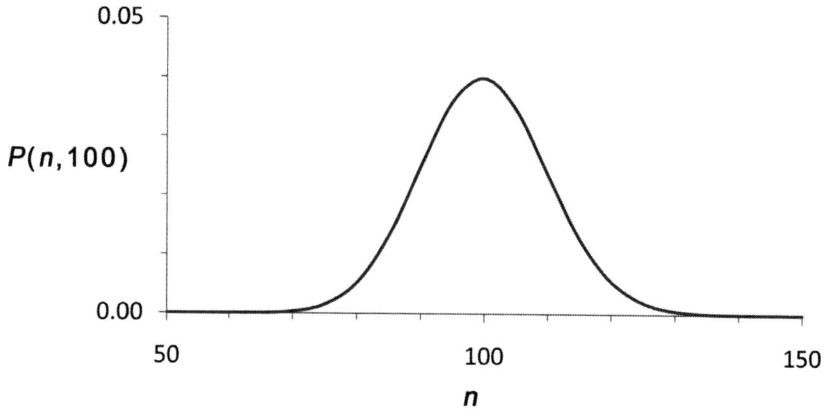

Figure 8.2 Poisson distribution for a mean quanta of 100.

$$\frac{\Phi}{\Phi_{SN}} = \Phi_{SN} = \sqrt{q_S Q} \quad \text{as } Q > 1000. \tag{8.6}$$

The total flux of the shot noise of the system is fairly estimated by a sum of the squares:

$$\Phi_S^2 = q_S \Phi_O + q_S \Phi_B + q_S \Phi_D. \tag{8.7}$$

Ideally, the shot noise of the object should be the limiting noise of the system. However, this is not always possible.

Chapter 19 provides more detail on shot noise.

8.4 Emittance Patterns

There are two common emittance patterns in microscopy: isotropic and Lambertian. An isotropic emittance pattern is uniformly distributed over a hemisphere. The radiance of an isotropic emittance is defined as

$$L_I = \frac{E}{2\pi}, \tag{8.8}$$

where E is the emittance. A radiance of a Lambertian object is dependent on $\cos\theta$:

$$L_L = \frac{E}{\pi} \cos\theta. \tag{8.9}$$

A Lambertian emittance is created by either a diffuse reflectance[10] or a specular transmission of isotropic emittance.[11]

8.5 Angular Collection Efficiency

The collection efficiency for a sphere of isotropic emission is

$$\eta_{SIE} = \frac{\Omega_C}{4\pi} = \sin^2 \frac{\theta_n}{2}, \tag{8.10}$$

where θ_n is an angle within the refractive medium of emission. In small-angle format, the marginal NA simplifies this expression:

$$\eta_{SIE} = \frac{\mathrm{NA}_M^2}{4n^2}. \tag{8.11}$$

The collection efficiency for a hemisphere of isotropic emission is

$$\eta_{HIE} = \frac{\Omega_C}{2\pi} = 2\sin^2 \frac{\theta_n}{2}. \tag{8.12}$$

In small-angle format, this becomes

$$\eta_{HIE} = \frac{\mathrm{NA}_M^2}{2n^2} = 2\eta_{SIE}. \tag{8.13}$$

The collection efficiency for a hemisphere of a Lambertian emission is derived by integration of a normalized Lambertian radiance as follows:

$$\eta_{HLE} = \int_0^{\theta_n} \left(\frac{\cos\theta}{\pi} \right) (2\pi \sin\theta \, d\theta) = \sin^2 \theta_n, \tag{8.14}$$

where the first group in parentheses is the normalized Lambertian radiance, and the second group in parentheses is the differential solid angle.

At small angles, the collection efficiency for a hemisphere of a Lambertian emission is

$$\eta_{HLE} = \frac{\mathrm{NA}_M^2}{n^2} = 2\eta_{HIE}. \tag{8.15}$$

The Lambertian emittance is more concentrated at small angles than an isotropic emission.

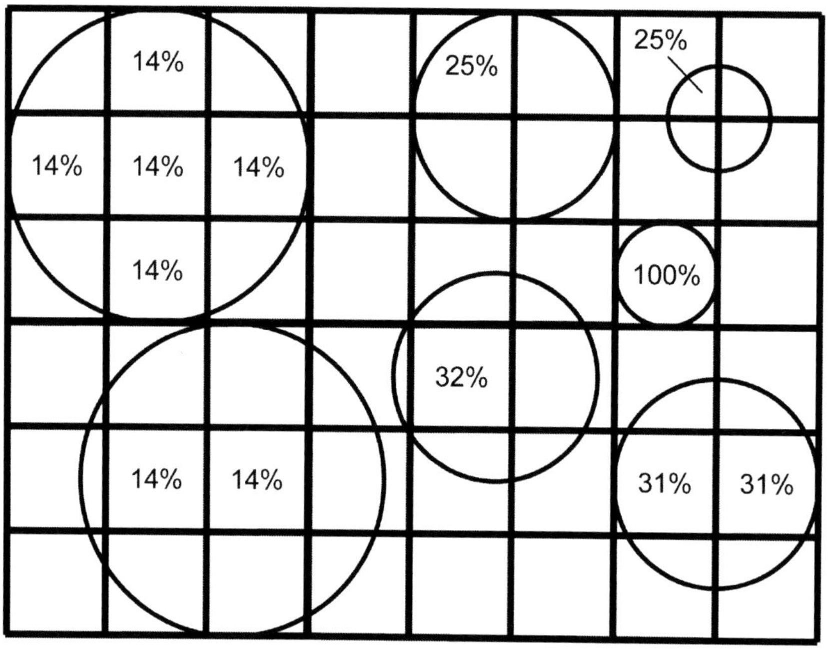

Figure 8.3 Spatial collection efficiency of point spreads on a pixel array.

8.6 Spatial Collection Efficiency

A point spread does not normally fit within a single pixel. A collection efficiency of the pixel η_P is defined by the portion of a point spread that overlaps a pixel. The maximum reliable collection by a pixel is 25%.

Figure 8.3 displays several point spreads within a pixel array. A point spread may fit within a pixel, as indicated by the 100% point spread. However, it is likely to split by 2 to 4 pixels. Consequently, the minimum reliable collection by a pixel is 25% for point spreads of a radius at pixel width or less. Reduction of the point spread does not increase the minimum reliable collection of pixel.

An optimum point spread fills four pixels. This yields a maximum signal strength of a point source object. They may also be combined into a super pixel of 2×2 pixels.

Image-processing algorithms benefit from contiguous full pixels. A point spread may fit within a 3×3 super pixel. This yields a consistent pixel collection efficiency of 14%, wherein at least one pixel is filled completely by the point spread. More pixels are definitely better for image processing; however, more pixels can yield smaller object flux per pixel while detector flux per pixel remains the same. The following expressions for contrast indicate the detriments of oversampling. The developer of an algorithm must consider the decreased contrast with background and noise that results from increased sampling of the object.

8.7 Full-Pixel Contrast

Étendue describes the space-angle product for rays. The term is derived from the French word for occupancy. In a geometric optical system, the etendue of a pixel is conserved from sensor to projected pixel:[10]

$$A_0\Omega_0 = A_1\Omega_1 = \cdots = A_n\Omega_n, \tag{8.16}$$

where A is the area of an image and Ω is the solid angle of the lens at the center of the image. The contrast of a full-pixel object[10] is expressed in large-angle format as follows:

$$C_{FPO} = \frac{L_O - L_B}{L_B + \Phi_D A_{PP}^{-1}\Omega_O^{-1}}. \tag{8.17}$$

The terms of pixel-based radiometry are defined Table 8.2. They are largely based on the projection of the pixel on the object.

Table 8.2 Terminology of pixel-based radiometry.

Symbol	Name
A_O	Area of the object
A_{PP}	Area of the projected pixel at the object
d_O	Dimension of the object
d_P	Dimension of the pixel at the detector
d_{PP}	Dimension of the projected pixel at the object
E_B	Emittance of the background
E_O	Emittance of the object
Φ_D	Flux of the detector
Φ_O	Flux of the object
η_C	Spherical collection efficiency of the central aperture
η_D	Quantum efficiency of the detector
η_M	Hemispherical collection efficiency of the marginal aperture
η_P	Efficiency of a pixel
η_{SR}	Efficiency of the Strehl ratio
L_B	Radiance of the background
L_O	Radiance of the object
NA_M	Marginal numerical aperture at the object
NA_C	Central numerical aperture at the object
n	Refractive index at the object
λ	Spatial wavelength within air
θ_M	Marginal angle of the lens stop at the object
θ_C	Central angle of the lens stop at the object
Ω_O	Angle of collection at the object

The small-angle format of contrast for a full-pixel object is

$$C_{FPO} = \frac{L_O - L_B}{L_B + \Phi_D\left(M^2 d_P^{-2}\right)\left(\pi^{-1}\mathrm{NA}_M^{-2} n^2\right)}. \tag{8.18}$$

The full-pixel contrast displays several dependences of interest. The effect of the detector is enhanced by the magnification and refractive index of the object. The effect of the detector is reduced by the dimension of the pixel and the marginal NA.

8.8 Subpixel Contrast

The contrast of a distant subpixel object is proportional to the square of the object dimension and the square of the resolution.[10] Figure 8.4 displays the redistribution of subpixel radiance within a full pixel. The large-angle format of subpixel contrast is adapted to a microscope as follows:

$$C_{SPO} = \frac{\left(L_O - L_B\right) A_O A_{PP}^{-1}}{L_B + \Phi_D A_{PP}^{-1} \Omega_O^{-1}} = \frac{A_O}{A_{PP}} C_{FPO}. \tag{8.19}$$

The small-angle format is

$$C_{SPO} = \frac{\left(L_O - L_B\right) d_O^2}{L_B\left(d_P^2 M^{-2}\right) + \Phi_D\left(\pi^{-1}\mathrm{NA}_M^{-2} n^2\right)}. \tag{8.20}$$

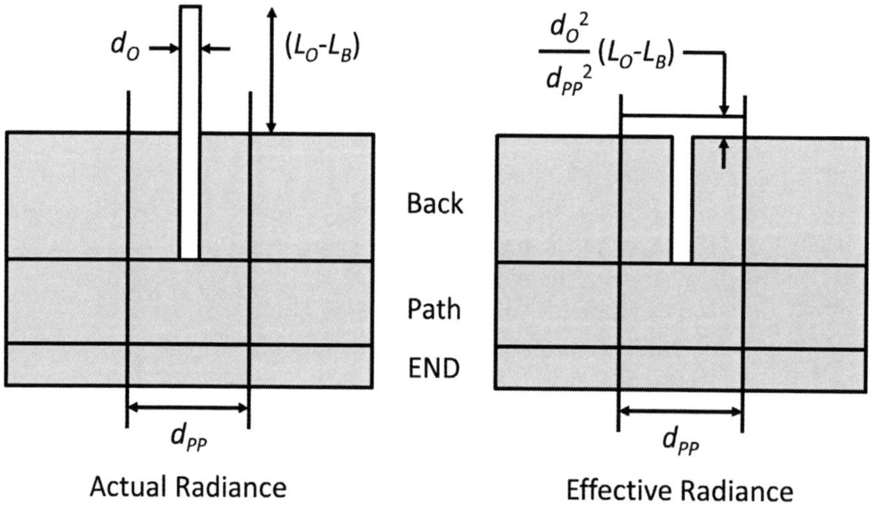

Figure 8.4 Redistribution of the radiance of a subpixel object. Contributing signals include the object (O), background (back), path, and equivalent noise of detector (END).

The subpixel contrast displays several dependences of interest. The effect of the background radiance is scaled by the area of the projected pixel $d_P^2 M^{-2}$. Consequently, the effect of the background radiance is enhanced by the dimension of the pixel and reduced by the magnification. The effect of the detector is reduced by the marginal NA and enhanced by the refractive index of the object.

The pixel collection efficiency in the above relationships apply to the geometric boundaries of a pixel. However, an optical system is rarely described by the geometric conditions. A geometric point expands into a spot. The radius of point spread normally equals a pixel width or greater. The diffraction-limited spot is specified by an Airy radius or a Gaussian radius, whereas the aberration-limited spot is specified by an rms radius or a geometric radius.

8.9 Point-Source Contrast

The contrast of a point-source object is a special case for a subpixel object. The geometric point source is transformed into a spot. An encircled-energy plot may calculate the pixel collection from the optical prescription. An MTF measurement may yield a central NA and a Gaussian NA.

The geometric point-source contrast is

$$C_{GPS} = \frac{\Phi_O \eta_P \eta_C}{L_B A_{PP} \Omega_O + \Phi_D}. \tag{8.21}$$

The small-angle format is

$$C_{GPS} = \frac{\Phi_O \eta_P \left(\pi\ \mathrm{NA}_C^2 n^{-2}\right)}{L_B \left(d_P^2 M^{-2}\right)\left(\pi\ \mathrm{NA}_M^2 n^{-2}\right) + \Phi_D}. \tag{8.22}$$

The geometric point-source contrast displays several dependences of interest. The effect of the background radiance is scaled by the area of the projected pixel. Consequently, the effect of the background radiance is enhanced by the dimension of the pixel and reduced by the magnification. The effect of the detector is reduced by the marginal NA and enhanced by the refractive index of the object.

An increased sampling of the point spread decreases the pixel collection efficiency. However, the above relationship indicates steady contributions from the background and the detector while the object signal is reduced by increased sampling. The developer of an algorithm must consider the reduction of object contrast when increasing the sampling of point spread.

8.10 Full-Pixel Airy Contrast

A special case is defined by the diffraction-limited emission of a fluorophore. The Airy pattern at the object is defined by the central NA. The spatial Airy pattern does not change between air and the immersion medium. However, the angle of collection does change between air and the immersion medium. The spherical collection efficiency (SCE) of a fluorophore is defined by the central NA within the refractive medium. The background fluorescence is isotropic within the immersion medium. The hemispherical collection efficiency (HCE) of the background is based on the marginal NA. These conditions define a large-angle format of the contrast for an Airy pattern as follows, with the help of Eqs. (5.20), (8.10), and (8.12):

$$C_{FPA} = \frac{E_O \eta_C}{E_B \eta_M + \Phi_D A_{PP}^{-1}} = \frac{\left(\Phi_O \pi \mathrm{NA}_C^2 \lambda^{-2}\right)\left[n^{-2}\sin^2(\theta_C/2)\right]}{E_B\left[2n^{-2}\sin^2(\theta_M/2)\right] + \Phi_D d_P^{-2} M^2}. \tag{8.23}$$

The small-angle format is

$$C_{FPA} = \frac{\left(\Phi_O \pi \mathrm{NA}_C^2 \lambda^{-2}\right)\left(\mathrm{NA}_C^2/4n^2\right)}{E_B\left(\mathrm{NA}_M^2/2n^2\right) + \Phi_D d_P^{-2} M^2} = \frac{\Phi_O\left(\pi \mathrm{NA}_C^4 \lambda^{-2} n^{-2}\right)}{E_B\left(2\mathrm{NA}_M^2 n^{-2}\right) + \Phi_D\left(4 d_P^{-2} M^2\right)}. \tag{8.24}$$

The full-pixel Airy contrast displays several dependences of interest. The effect of the detector increases with magnification. The object signal grows faster than the background until NA_M exceeds NA_C. However, as NA_M exceeds NA_C, the object signal remains constant while the background signal grows rapidly with NA_M. Ideally, NA_M should not exceed NA_C. Also, a higher sampling of the Airy pattern through a larger magnification degrades contrast of the object with the detector.

The marginal and central NAs are related by the efficiency of the Strehl ratio as follows:

$$\eta_{SR} \mathrm{NA}_M^2 = \mathrm{NA}_C^2. \tag{8.25}$$

The full-pixel Airy contrast is expressed in terms of the Strehl ratio as follows:

$$C_{FPA} = \frac{\Phi_O\left(\eta_{SR}^2 \pi \mathrm{NA}_M^4 \lambda^{-2} n^{-2}\right)}{E_B\left(2\mathrm{NA}_M^2 n^{-2}\right) + \Phi_D\left(4 d_P^{-2} M^2\right)} = \frac{\Phi_O\left(\eta_{SR}^2 \pi \lambda^{-2} \mathrm{NA}_M^2 2^{-1}\right)}{E_B + \Phi_D\left(2 d_P^{-2} M^2 \mathrm{NA}_M^{-2} n^2\right)}. \tag{8.26}$$

The full-pixel Airy contrast increases rapidly with marginal NA until the Strehl ratio starts dropping below unity.

A developer of an algorithm must consider the detrimental effects of increased sampling on the Airy pattern. The required magnification increases the effect of detector noise. Also, the increased marginal NA of the larger magnification does not collect more object signal, whereas it does collect more background. Thus, an increased sampling reduces contrast when the Strehl ratio is reduced.

Chapter 9
Microlens Formats

Most camera lenses are designed for a range of 500 mm to ∞. This range accommodates both a head shot and a landscape. A typical camera lens for a 35-mm-film format is 52 mm in focal length. According to Eq. (3.7), a magnification of 0.1 requires an object distance of $11f$, or 550 mm. This may also be considered a 10X reduction. The performance of a typical camera lens degrades rapidly beyond a 0 to 10X reduction range.

The terms macrolens and microlens are interchangeable. In photography, a macrolens converts a small object into a large image within a print. In microscopy, a microlens looks at a small object. They both operate within a reduction range of 10XR to 1X. A magnification range of 1X to 10X is achieved by simply reversing the lens. Nikon frequently employs microlens as the name for a lens with near-unity magnification.

9.1 10XR Double Gauss

Figure 9.1 displays a 10XR double-Gauss lens for application to a film format with a 35-mm diagonal. The lens, as described in Smith,[12] indicates a large spherical aberration and a decreasing relative illumination with field height. The decreased relative illumination indicates *f*/# inflation with field angle. Rays of the off-axis field must be clipped to manage astigmatism. Inflation of *f*/# with field angle is a characteristic of a double-Gauss lens. The actual field is adapted to a sensor with a 12-mm diagonal, or 6.0-mm field height, or 6.9-deg field angle. The *f*/# inflation at 6.9 deg is < 1%.

The ray-intercept plot indicates enormous aberration at a 6.0-mm field height. The tangential error indicates 154 μm of spherical aberration. The sagittal error indicates 51 μm of spherical aberration. The edge spread is nearly 40 μm. The encircled energy at 10 μm is much less than 30%. The flat spot of the ensquared plot occurs at much less than 5 μm. The diffraction limit might be suitable for a 10-μm pixel, while the aberration limit is certainly not.

The low *f*/# indicates a high brightness; however, the aberrations indicate poor contrast. A full-pixel object at > 40 μm in width should display a brightness in accord with the *f*/#. However, edges will spread to > 20 μm wide. A 40-μm-wide stripe transforms into a round peak at 80 μm in width. A subpixel object will spread into a mound of half height at 40 μm in diameter.

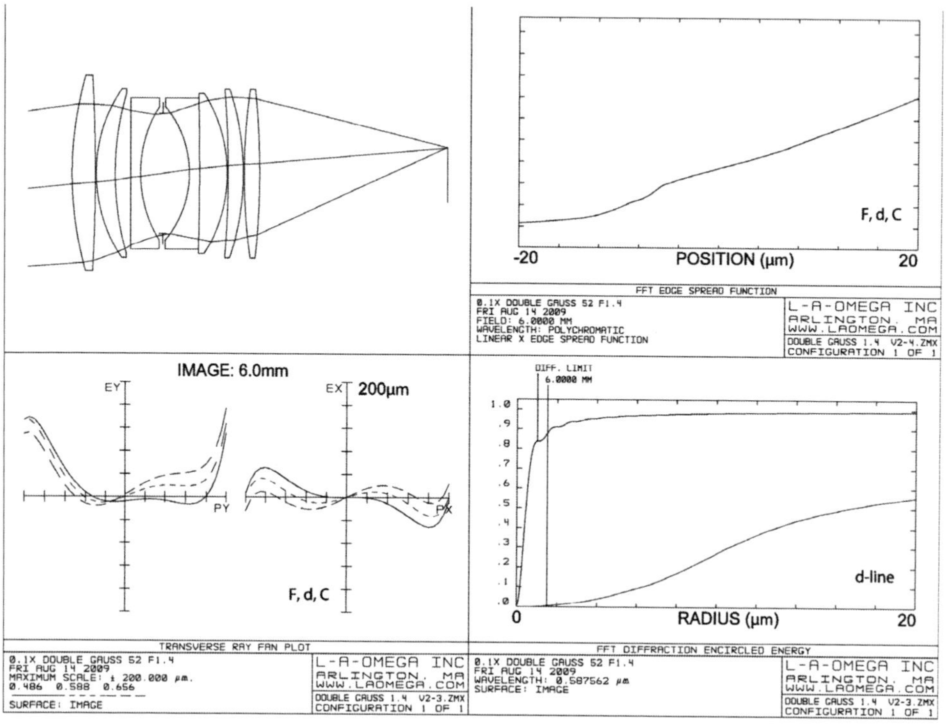

Figure 9.1 10XR double-Gauss F/1.4. Focal length is 52 mm. Lens *f/#* is 1.4. Image space *f/#* is 1.5. Image NA is 0.31. Edge spread: 512 × 512 sampling. Encircled energy: 512 × 512 sampling. Prescription by G. H. Smith (2009), Lens 23.3.[12]

9.2 10XR Microlens

Figure 9.2 displays a microlens at 10XR[13] for application to an image sensor with a 12-mm diagonal of 6.0-mm image field height. The sagittal plane indicates very little spherical aberration, as the plots are nearly linear. The C, d, and F lines are not well corrected to each other. The tangential plane indicates a small amount of coma in the negative portion of the pupil. Overall, this lens functions quite well at 10XR. The image-space *f/#* is 3.2.

The microlens at 10XR functions effectively for charge-coupled device (CCD) sensors. The edge spread is approximately 10 μm, which is on the order of most CCD pixels but not CMOS pixels. The flat of the encircled plot indicates a diffraction limit of approximately 2.2 μm in radius. The aberration limit is approximately 7 μm, which is again appropriate for most CCD pixels. At 10XR, this is a terrific lens for CCD sensors with 8- to 10-μm pixels. However, there is significant axial color as indicted by the broad line spread.

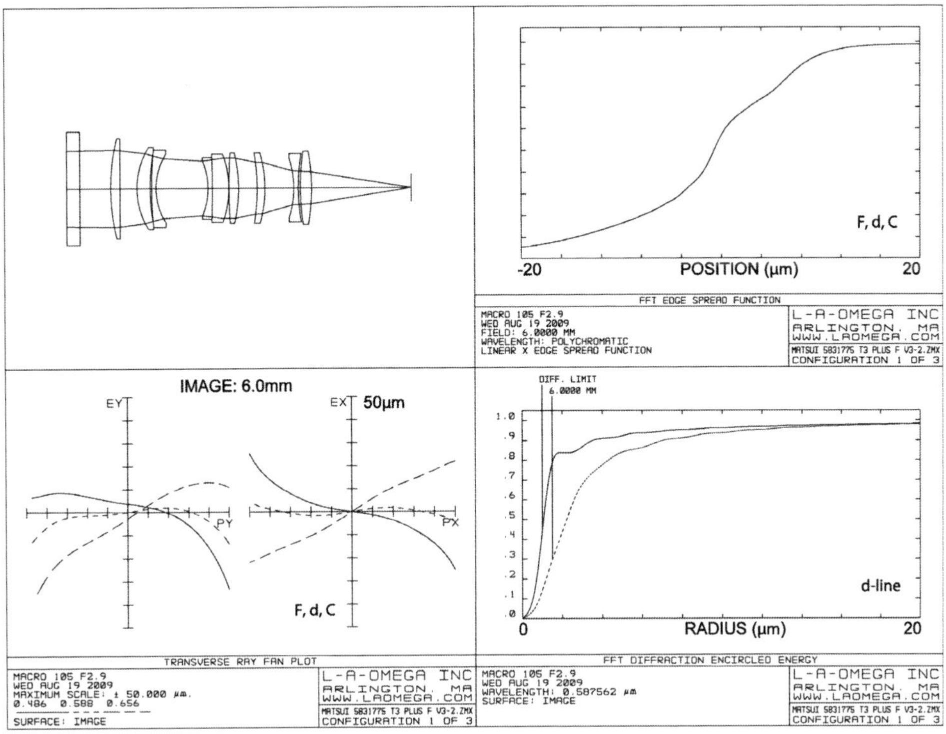

Figure 9.2 10XR microlens F/2.9. Focal length is 100 mm. Image space *f/#* is 3.2. Image NA is 0.15. Edge spread: 64 × 64 sampling. Encircled energy: edge: 64 × 64 sampling. Prescription by Matsui.[13] Flat plate added as a filter in object space.

9.3 2XR Microlens

Figure 9.3 displays a microlens at 2XR for application to an image sensor with a 12-mm diagonal of 6.0-mm image field height. The sagittal plane indicates both third- and fifth-order spherical aberration. The C, d, and F lines are not well corrected to each other. The tangential plane indicates a small amount of lateral color for the F line (486 nm), as it shifts upward by 5 μm. The tangential plane indicates a small amount of astigmatism for the F line, as it slopes upward. The image-space *f/#* is 4.3.

The microlens at 2XR functions adequately for CCD sensors. The edge spread is approximately 16 μm, which is nearly 2 pixels of a CCD pixel. The flat of the encircled plot indicates a diffraction limit of approximately 3.0 μm in radius. The aberration limit is approximately 14 μm, which is nearly 2 pixels of a CCD pixel. At 2XR, this lens is limited by aberration beyond a CCD pixel size. It is marginally optimized for 10 μm without manufacturing considerations. A tolerance budget during manufacture would indicate larger errors. There is significant axial color as indicted by the broad line spread.

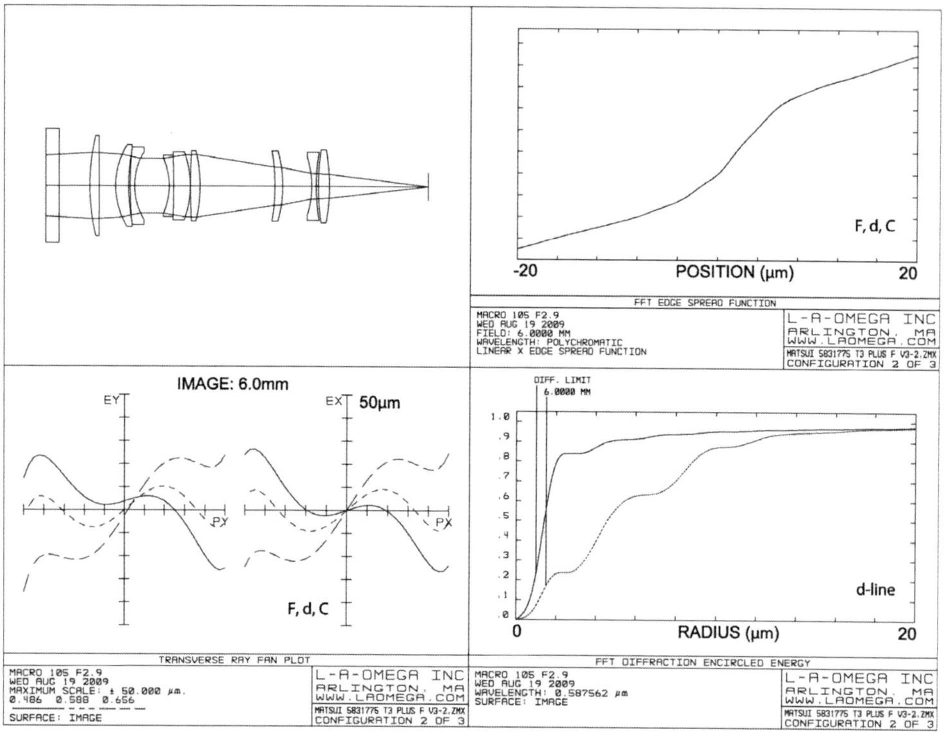

Figure 9.3 2XR microlens F/2.9. Focal length is 100 mm. Image space *f/#* is 4.3. Object NA is 0.058. Image NA is 0.116. Edge spread: d line, image field 6.0 mm, 64 × 64 sampling. Encircled energy: edge: d line, image field 6.0 mm, 64 × 64 sampling. Prescription by Matsui.[13] Flat plate added as a filter in object space.

9.4 1X Microlens

Figure 9.4 displays a microlens at 1X for application to an image sensor with a 12-mm diagonal of 6.0-mm image field height. The sagittal plane indicates both third- and fifth-order spherical aberrations, where the fifth-order aberration is larger than the third. The C, d, and F lines are not well corrected to each other. The tangential plane indicates a small amount of lateral color for the F line and a small amount of coma for all wavelengths, as the plots all curve downward. The image-space *f/#* is 5.5.

The microlens at 1X functions poorly for CCD sensors. It has similar problems as the microlens at 2XR but with larger magnitude. The edge spread is > 20 µm. The flat of the encircled plot indicates a diffraction limit of approximately 4.0 µm in radius. The aberration limit is > 20 µm. The microlens at 1X aberration beyond a CCD pixel size is not optimized for a 10-µm pixel. A tolerance budget in manufacture would indicate larger errors.

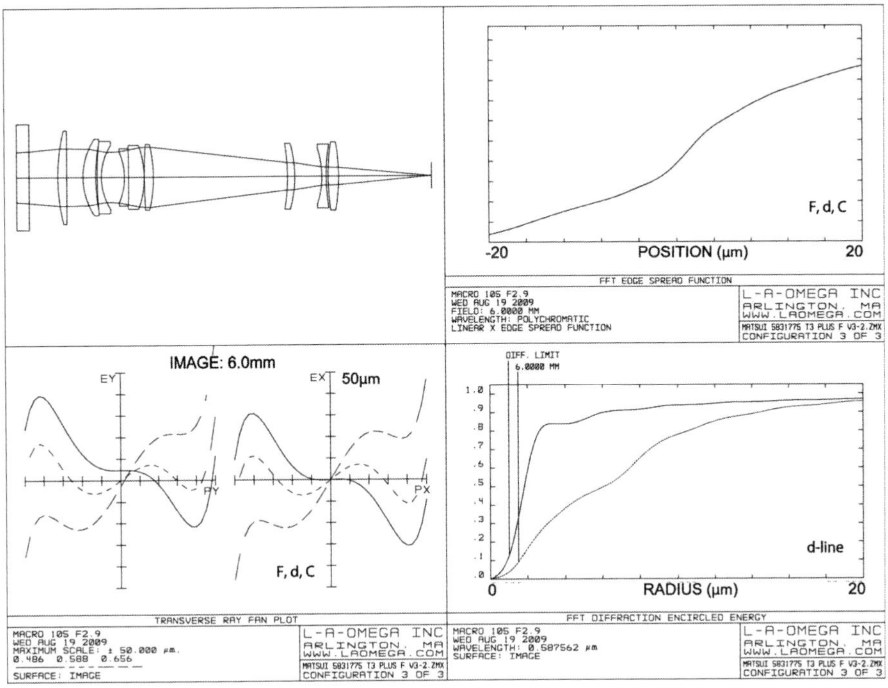

Figure 9.4 1X microlens F/2.9. Focal length is 100 mm. Image space *f/#* is 5.5. Object NA is 0.089. Image NA is 0.089. Edge spread: 64 × 64 sampling. Encircled energy: edge: 64 × 64 sampling. Prescription by Matsui.[13] Flat plate was added as a filter in object space.

9.5 2XR Telecentric Spectroscopy Lens

Figure 9.5 displays a telecentric spectroscopy lens (TSL)[14] at 2XR for application to an image sensor with a 12-mm diagonal of 6.0-mm image field height. The sagittal plane indicates no spherical aberration at the d line. The reduced spectrum ends at the Argon line (514 nm), which displays a small third-order spherical aberration. The C line (656 nm) displays an 8-μm third-order spherical aberration. The argon, d, and C lines are corrected within a small spherical aberration. The spherochromatism is well balanced for the three wavelengths. The TSL at 2XR is not an achromat, but it might be called a semiapochromat because it is nearly corrected for three wavelengths. The tangential plane indicates a small amount of coma at 8 μm in the positive portion of the pupil. The image-space *f/#* is 4.0.

The TSL at 2XR functions effectively for CCD sensors. The edge spread rises sharply over 4 μm for ease of edge of detection by the image-processing algorithm. The total width of the edge spread is about 10 μm, which is nearly 1 pixel. The flat of the encircled plot indicates a diffraction limit of approximately 3.0 μm in radius. The aberration limit is < 10 μm for most of the rays. At 2XR this lens is limited by aberration at a typical CCD pixel size of 10 μm. It is optimized for 10 μm with consideration of manufacture. A reasonable tolerance budget of 0.1-mm decenter in manufacture does not indicate larger errors.

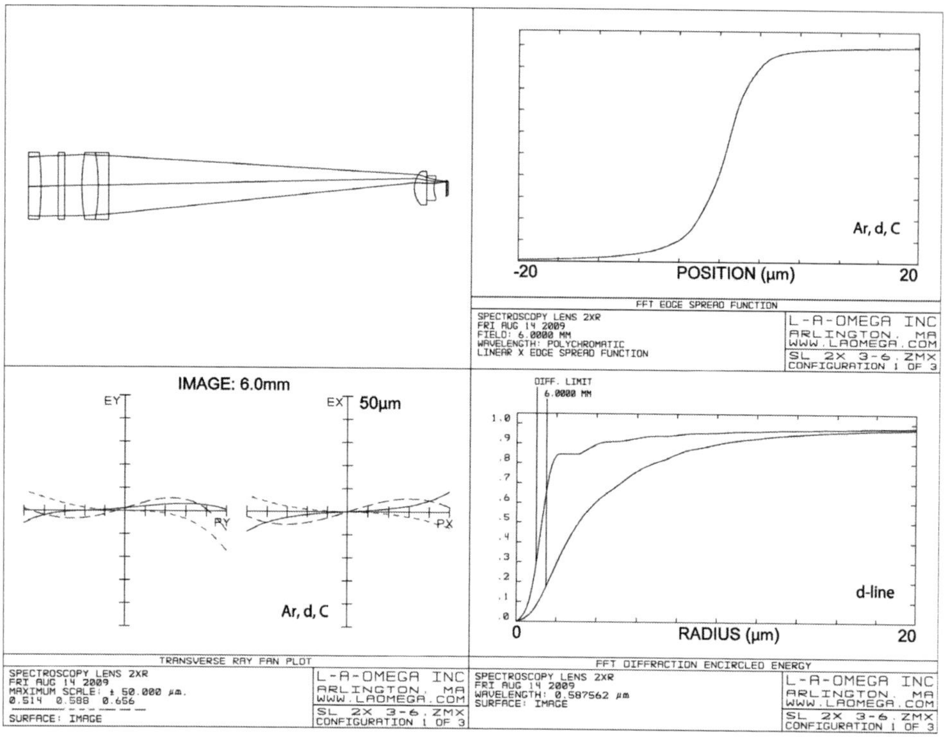

Figure 9.5 2XR telecentric spectroscopy lens. Wavelengths: Ar (argon 514 nm), d line (588 nm), and C line (656 nm). Focal length is 193 mm. Image space *f/#* is 4.0. Object NA is 0.063. Image NA is 0.126. Edge spread: d line, image field 6.0 mm, 64 × 64 sampling. Encircled energy: edge: d line, image field 6.0 mm, 64 × 64 sampling. Prescription patent pending by Seward.[14]

Figure 9.6 displays the two groups of the TSL. Group 1 corrects the primary aberrations, which are largely dependent on NA. The aspheric surface of group 1 manages the third- and fifth-order spherical aberration. The doublet manages the axial color and coma. Group 2 corrects the secondary aberrations of field angle. The aspheric surface of group 2 manages astigmatism and field curvature. The concave surface also manages astigmatism and field curvature. A lanthanum krone (LaK) of the aspheric lens of group 2 manages lateral color. The flange of the asphere of group 2 provides the precise axial location of group 2. The chief rays are telecentric, which is beneficial for sensors with a microlens over each pixel. This lens is highly optimized for spectroscopy. A filter is placed in the infinity correction zone of group 1. Consequently, filters can be swapped without disruption of focus. The reduced spectrum of 514–656 nm eliminates chromatic aberrations of the blue spectrum, which are frequently not active in spectroscopy.

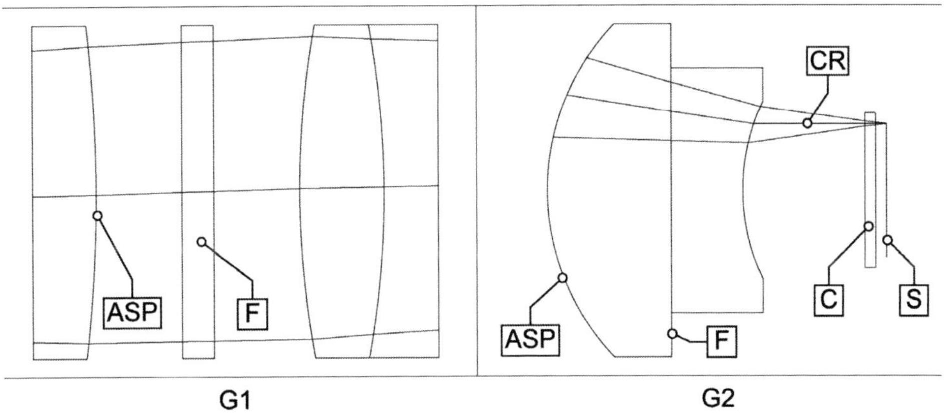

Figure 9.6 Lens groups of 2XR telecentric spectroscopy lens. Aspheric surface (ASP), filter of group 1 (F), flange of group 2 (F), chief ray (CR), cover glass (C), and sensor (S).

The object space defines a 0.063 NA. Application of Eq. (1.8) to the d line yields a Gaussian depth of 94 μm. A 0.19-mm depth of focus is reasonable for consistent operation. A reasonable tolerance is essential for automated loading as well as accommodation of tilt of the specimen plate. Effective management of tolerances is a critical part of instrument design.

The hemispherical collection efficiency (HCE) of the TSL is displayed in Fig. 9.7. The object space of the marginal NAs are calculated from the marginal ray angles. The 2XR TSL (0.063) displays a total HCE as 0.000 55. The 2XR ML (0.058) displays a total HCE as 0.000 46. The 1X ML (0.089) displays a total HCE as 0.001 07.

The ML at 1X is a fine lens for imaging of fluorescent arrays. The 1X ML collects nearly twice as much light as the TSL and ML at 2XR. However, the maximum HCE of the 1XR requires a 20-μm pixel. The higher object NA defines a smaller depth of focus, which is difficult to manage by tolerance of placement.

The TSL offers superior color correction to both versions of the ML. A filter may be swapped without a disruption of focus. The fewer pieces define a simpler tolerance budget and better transmittance. However, a disadvantage of the TSL is cost.

An aspheric lens in small quantities is expensive to manufacture. A coordinated process of interferometry, grinding, and polishing is required. An aspheric lens with a 60-mm diameter and a wavefront error of 0.1λ costs about $3000 per unit in the year 2009. Fortunately, this cost may be recovered through a reduction in consumption of chemicals during operation.

The TSL is suited for applications wherein photons are precious. Fluorophores can photobleach after just 200 photons of emission, reagents can be expensive, and potential drug compounds from extinct plants may be used only in small amounts. There are many applications for the telecentric spectroscopy, even with a 10X cost over a mass-production lens.

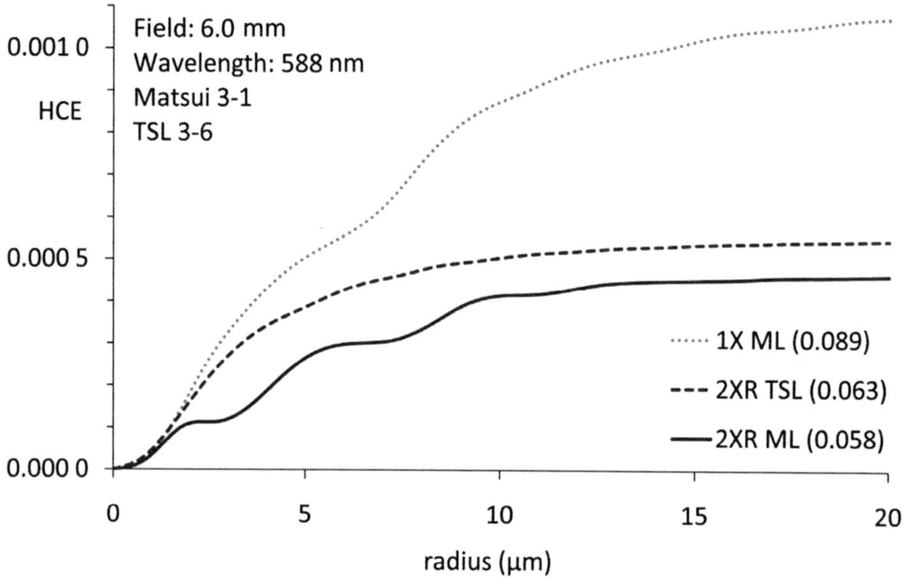

Figure 9.7 Hemispherical collection efficiency (HCE) of a telecentric spectroscopy lens (TSL) and a microlens (ML). The HCE is plotted versus radial position within the point spread. The HCE plot is an encircled-energy plot from ZEMAX that is scaled by the HCE. The encircled-energy plot occurs in air (n = 1.00). HCE occurs in water (n = 1.33). Marginal NAs of object space are indicated within parentheses.

Chapter 10
Illumination Systems

10.1 Condenser

As shown in Fig. 10.1, a condenser is a relay lens for the field stop of an illumination system. The illumination field stop is a real object. The condenser creates a virtual image of the illumination field stop at the specimen. The illumination field should be slightly larger than the vision field. Assembly tolerances dictate the amount of additional size of the illumination field beyond the vision field. The condenser also creates a virtual image of the illumination lens stop at the vision lens stop.

10.2 Abbe Illumination

Abbe illumination locates a source at a conjugate of the illumination field with a small NA. Earnst Abbe promoted a small NA of illumination, which facilitates the Abbe sine condition.[15] The small illumination NA of Abbe may be defined as < 0.75 times the vision NA.

Earnst Abbe worked for Carl Zeiss. According to the Zeiss archives, Abbe eventually became the chief executive at Zeiss Optics, through which position he stimulated growth of the company from 25 to 1,400 employees. Abbe promoted numerous social reforms at Zeiss Optics, such as paid vacation, profit sharing, pensions, sick time, and an eight-hour workday.

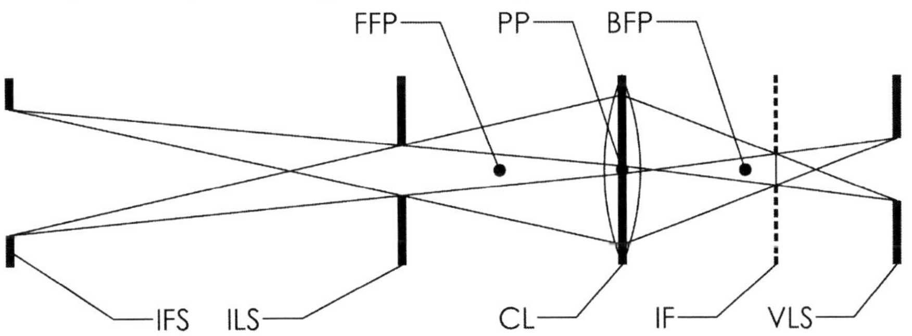

Figure 10.1 Condenser lens (CL), illumination field (IF), and stops: illumination field stop (IFS), illumination lens stop (ILS), and vision lens stop (VLS). Cardinal points: Front focal point (FFP), principal point (PP), and back focal point (BFP).

Abbe is frequently credited with many advancements of optics in the late 1800s. He defined the term "numerical aperture" in 1878,[16] which replaced terms such as "water angle" and "balsam angle." However, he was not responsible for "critical illumination," which is properly credited to Nelson in the next section.

10.3 Nelson Illumination

Nelson illumination locates a source at a conjugate of the illumination field with a large NA. The critical illumination NA of Nelson may be defined as ≥ 0.75 times the vision NA. Nelson illumination originally applied to the flame of an oil lamp as the light source. The isotropic radiance of the flame could be placed at a field conjugate without a diffuser. The extent of the source defines the illumination field stop. A large angular size of the lens stop creates a small radius of partial coherence at the illumination field.

In Fig. 10.2, the image of a source is relayed onto the illumination lens stop. A source collector lens gathers as much light as is practical. A source field lens is located near the illumination field stop. The source field lens directs light of the field stop into the lens stop.

In 1875, Edward Nelson discovered "the advantage of a large axial cone."[15] However, his contradiction of small-cone illumination created trouble for him. In 1889, Abbe openly condemned the large-angle cone.[15] Abbe also denounced a large annulus by Smith.[15] Eventually, after repeated exhibitions by the Royal Society of Microscopy and others, the large-angle cone of Nelson was adopted in high-class microphotography. Smith's annular illumination was also proven through photographic evidence. In 1910 Nelson documented his critical illumination, 35 years after his discovery and 21 years after its public condemnation. Nelson defined critical illumination as "the aperture [of illumination]...is not less than three-quarters of the NA of the observing objective."[15]

Critical illumination is beneficial for maximum separation of two overlapping point spreads. However, it is not required for detection of incoherent fluorescent emissions. Critical illumination minimizes the coherent interaction between point spreads.

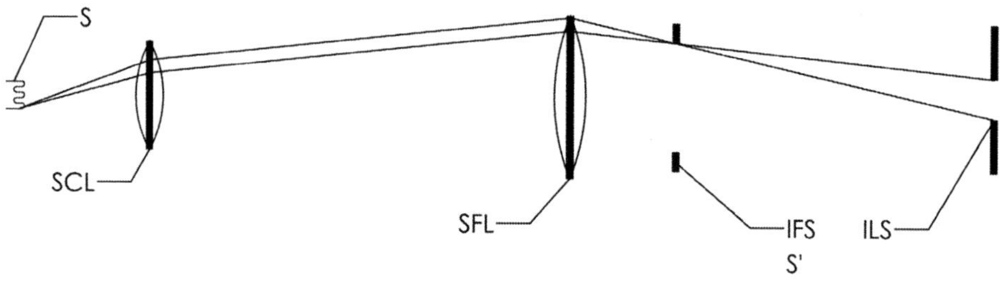

Figure 10.2 Source relay of critical illumination. Source (S), source collector lens (SCL), source field lens (SFL), source image (S'), illumination field stop (IFS), and illumination lens stop (ILS). Illumination NA should be 0.75 times vision NA.

The Abbe resolution[i] of Nelson illumination may be defined as

$$d_{Abbe} = \frac{\lambda}{\mathrm{NA}_V + \mathrm{NA}_S}, \text{ where } \mathrm{NA}_S \le \mathrm{NA}_V, \tag{10.1}$$

where d_{Abbe} is the minimum separation of the features, NA_V is the NA of the vision system, and NA_S is the NA of the source. A point source defines a completely coherent illumination field. The Abbe resolution of a coherent field is λ/NA_V. A broad source $\mathrm{NA}_S \ge 0.75\ \mathrm{NA}_V$ defines an incoherent system. The Abbe resolution of a broad source is limited by the vision system. The effective NA of the illumination field cannot exceed that of the vision field. Consequently, the Abbe resolution of an incoherent system cannot exceed $0.5\lambda/\mathrm{NA}_V$. The partial coherence of the illumination resembles the point spread of an Airy pattern as follows:

$$\Gamma(\rho) = \frac{2J_1(\rho\, k\mathrm{NA}_S)}{\rho\, k\mathrm{NA}_S}. \tag{10.2}$$

10.4 Diffusers

The coil structure of a filament is extremely nonuniform. Consequently, placement of a filament near an image conjugate creates an image of the filament at the sensor. A diffuser is required for reduction of the filament structure of the image.

A diffuser spreads light over both space and angle. Consequently, the etendue of the illumination increases and the radiance decreases. Increased uniformity by diffusion requires significant reduction of radiance. Consequently, the image brightness decreases. An improved uniformity over space and angle normally requires a reduction of image brightness.

Diffusers are notorious for inconsistent performance. The effects of an acid etch are dependent on time of exposure and composition of materials. Sand blasting employs variable grain size. Opal glass employs diffuse scattering throughout its volume. An opal diffuser creates an extremely isotropic radiance in transmission. However, much incident light is backscattered. The inconsistency of a diffuser in production might create enormous variations between instruments.

The emission of a filament is uniform over angle. Consequently, conversion of the angular extent of a filament into a spatial extent is highly beneficial. Ideally, the filament should be located at an image conjugate of the lens stop. This is a foundation of Köhler illumination.

[i] The Abbe resolution applies to the spacing of a grating.

10.5 Köhler Illumination

In 1893, Köhler conceived an illumination system where the flame of an oil lamp was focused on the lens stop of a relay lens.[17] Köhler stated two important principles, as translated in the English reprint on the centennial anniversary of his original article in German.[18] The first principle of Köhler illumination is: "the light source is almost in the back focal plane of the condenser lens."[18] This ensures sufficient defocus of the light source at the vision field. The image of the source is distant from the condenser lens. The second principle of Köhler is: "the location of the object...is beyond the focal length of the condenser."[18] This places the illumination field stop at a different location from the image of the source. Köhler also indicated an image of the illumination lens stop at the vision lens stop.

A modern interpretation of Köhler is based on three principles: the illumination lens stop is located at the front focal point of the condenser; an image of the source is located at the illumination lens stop; and the vision lens stop is located at an image of the illumination lens stop. The first condition ensures maximum defocus of the filament. The second and third conditions minimize reduction of image brightness at the field margin. These conditions can be mutually exclusive because most objective lenses do not have the vision lens stop located precisely at a focal point of the objective. The telecentric configuration of the first condition is most common; however, it can create an illumination profile with a significant dependence on field height.

In a Köhler illumination scheme, the field stop collects light from a filament at sufficient distance for excellent uniformity in the illumination field. The uniformity within the field stop is determined by the angular distribution of the filament's emission. The spatial distribution of the filament is located at the lens stop.

In Fig. 10.3, the filament relay of a Köhler system is displayed. The filament is relayed by the source collector lens on the illumination lens stop. An image of the filament is located on the illumination lens stop and the vision lens stop. The angular distribution of the source is converted by the source collector lens into a spatial distribution at the illumination field stop. Consequently, the spatial uniformity is excellent without any use of a diffuser. The illumination field stop is relayed onto the illumination field by the condenser lens.

The position of the filament is critical to Köhler illumination. Fabrication tolerances cannot reliably place an image of the filament within the lens stop. Consequently, the filament must be actively aligned, or the filament emission must be diffused. A typical filament has an operational lifetime of 1000 h.

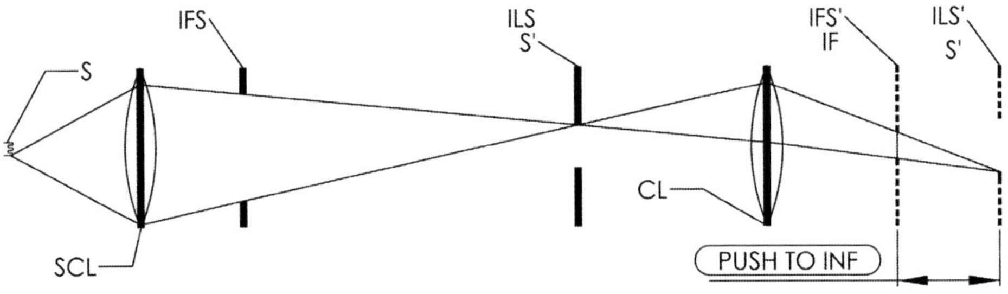

Figure 10.3 Principles of Köhler illumination. The image of a source (S′) is distant from the illumination field (IF). The image of an illumination lens stop (ILS′) is distant from the illumination field (IF).

The stops of Köhler illumination are displayed in Fig. 10.4. A first variable iris is placed at the illumination field stop. The illumination field stop defines an illumination field that should overfill the vision field. Illumination far beyond the vision field is detrimental to contrast because the specimen might scatter light in the vision field. A second variable iris is placed at the illumination lens stop. The illumination lens stop defines an illumination NA which should underfill the vision lens stop. Illumination beyond the vision lens stop is highly detrimental to contrast because the objective barrel is normally stainless steel, which reflects extraneous light into the vision system.

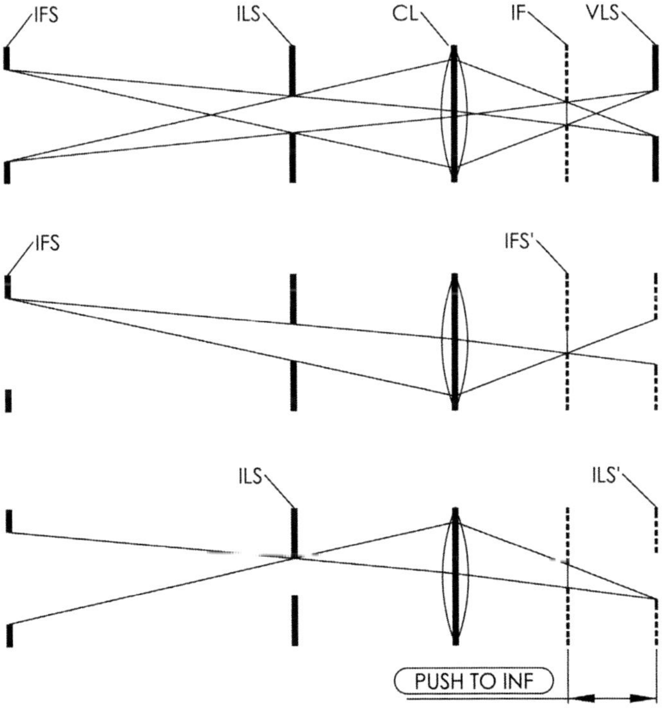

Figure 10.4 Stops of Köhler illumination: (a) maximum extent of rays, (b) relay of the illumination field stop (IFS), and (c) relay of the illumination lens stop (ILS).

An iris may also be called a diaphragm, which a is flexible membrane that covers on opening; however, a diaphragm does not normally have a hole in the middle as an iris does. A diaphragm of the lung controls breathing, a diaphragm of the ear converts air pressure into nerve signals, and a diaphragm of a speaker converts an electrical signal into air pressure. The word iris is derived from the Greek *iris*, meaning rainbow. Interestingly, a rainbow is a naturally occurring annulus that does resemble the iris of an eye, which is an optical stop.

At the limit of a point source, $NA_S = 0$ in Eq. (10.2), a Köhler illumination field becomes completely coherent, which is detrimental to image quality. Constructive interference between image point spreads normally creates detrimental features such as fringes. A laser beam resembles a collimated point source at $NA_S = 0$. The axial coherence of a monochromatic laser [Eq. (5.31)] creates both circular and linear fringes within an image. The small angular size of a Köhler source can create similar fringes over a short range of lateral coherence [Eq. (10.2)].

10.6 Matched Stops

In both Nelson and Köhler illumination, the space-angle product of the illumination is maintained by two variable irises. A first iris is placed at the illumination field stop, and the second is placed at the illumination lens stop. The space-angle product of the illumination should fill the objective lens stop and the vision field without overfilling either.

Overfilling the space-angle product of the vision system creates extraneous light in the background of the image. Illumination beyond the vision field may be scattered by the specimen into the vision system. Illumination beyond the vision lens stop may be scattered by the lens barrel into the vision system. A major benefit of Köhler illumination is maximum image brightness without reduction of contrast. The stops are adjusted to maintain the space-angle product without overfilling either the vision field stop or vision lens stop.

10.7 Light-Emitting Diodes

A light-emitting diode (LED) offers several advantages over an incandescent filament. The lifetime of an LED can reach 100,000 h, while an incandescent halogen bulb can only reach 2,000 h. LEDs can offer a spatially uniform emitter for critical illumination. Their angular emittance pattern is typically Lambertian. An isotropic internal emission is converted to an external Lambertian profile by Snell's law.[11] The peak of a Lambertian radiance profile is twice that of an isotropic profile; thus, more of the light is directed into a lens stop.

LEDs do, however, have disadvantages. An LED requires a binary semiconductor. Silicon and germanium are elementary semiconductors with indirect bandgaps, which cannot offer direct transition to an open state of the valance band. Gallium arsenide and other binary semiconductors have direct bandgaps, which offer transition to the open states of the valence band. Consequently, an LED requires a binary compound.

The radiance of the sun is eye safe at 0.5 deg but is not eye safe after magnification. The irradiance of the sun's image at the retina does not change after magnification; however, the cooling geometry of the image is much different. During normal vision, the 0.5-deg image of the sun is absorbed over a small sphere that creates a spherical cooling geometry, which permits safe dissipation of heat. During magnification, however, the image of the sun is absorbed over a disk that creates a planar cooling geometry, which cannot dissipate heat quickly enough to prevent damage to the eye. Consequently, any magnification of the sun might create a retinal burn.

This warning about the sun also applies to high-brightness LEDs. The American National Standards Institute (ANSI) has specific information on diffuse emitters, such as LEDs, as part of its standard for safe use of lasers, Z136.1.[19]

The eye has two defense mechanisms against bright light: a blink and a large saccade. A blink is the closure of the lid. The typical reaction time for a blink is 0.18 s. The ANSI standards[19] define an aversion response of 0.25 s for visible light; this defines a 5-mW limit for a class III(a) laser, which is safe within a blink of an eye. On the other hand, a class III(b) laser is not safe within a blink of an eye. A large saccade is a gaze shift that normally occurs within 10 s. The ANSI standards[19] define an aversion response of 10 s for invisible light.

Figure 12.14 in Schubert[11] indicates emission by material type. Gallium nitride with indium (GaInN) emits in the blue (450–490 nm) and green (515–570 nm) wavelength ranges. GaInN can be grown on a sapphire (Al_2O_3) substrate; however, there is significant strain due to lattice mismatch. Gallium phosphide (GaP) can emit weakly in the yellow (570–580 nm), which is near the peak of human sensitivity at 555 nm. Gallium arsenide with aluminum and indium (AlGaInAs) emits strongly in the orange (585–620 nm). Aluminum gallium arsenide (AlGaAs) emits strongly in the red (625–740 nm). There is a significant dead spot in LED emission in the green side of the yellow wavelength range (530–580 nm). LEDs offer much longer lifetimes than incandescent filaments. Schubert[11] provides a practical review of photopic vision, which is based on a red, green, and blue cones.

The peak wavelength of emission is strongly dependent on concentrations of the components during vacuum deposition. Typically, the peak wavelength varies by as much as the full width at half maximum of emission. Many dyes and fluorophores are dependent on a narrow band of excitation. A tolerance for peak wavelength of excitation should be established. The peak wavelength varies greatly by production run; a typical variation is 3–5% of the nominal peak wavelength.

A white LED is generated by a blue LED and white phosphor. The phosphor creates a uniform spectrum distribution along with a variable blue peak. A white LED has an artificial appearance in comparison to an incandescent bulb.

The packaging of an LED is dependent on its power. A low-power package encapsulates the die and leads in epoxy, which also provides structural support. A high-power LED package requires a heat sink in direct contact with the LED chip. The LED chip of a high-power package can be exposed to the air or covered by a lens. A high-power LED package typically employs silicone as the encapsulate.

Epoxy resin provides excellent stability over time. However, an epoxy resin cannot tolerate high temperatures (> 120° C). Epoxy might also exhibit reduced transmission of red, violet, and UV rays. The refractive index of epoxy is typically 1.6, while the refractive index of GaP is approximately 3.5. A higher refractive index of the encapsulate promotes extraction of photons from the LED chip. A low-power package frequently employs an epoxy lens at a 5-mm-diameter T1-3/4 package. The placement error of the chip within a T1-3/4 package is large, as the decenter can exceed the emitter width.

A silicone encapsulate can tolerate the higher temperatures (190° C) of high-brightness LEDs. Polysiloxane is the correct chemical name, which indicates a chain of siloxane groups R_2SiO, with a variety of side groups R. The refractive index can range from 1.4 to 1.6 with dependency on the composition of the side groups R. The stiffness is also dependent on the composition of the side groups. A high-brightness LED typically employs a 5.6-mm-diameter hemispherical dome as the encapsulate.

There are numerous lenses that fit over the standard 5.6-mm dome. An LED lens typically comprises a central refractor and an annular reflector.[20] The combination of refractive and reflective elements defines a catadioptric LED lens. The material normally used is polymethylmethacrylate (PMMA), with a refractive index of 1.49. The uniformity of a catadioptric LED lens is not suitable for application to microscopes.

10.8 Aspheric Plus Singlet Relay

Figure 10.5 displays a lens report for a condenser comprising an aspheric lens and a singlet. They are both stock items as indicated in the prescription. A small amount of coma limits the d line to a sharp focus of 10 μm. The axial color limits the polychromatic focus to 160 μm.

A magenta ring is present at the perimeter of the illumination field. The polychromatic edge spread indicates two distinct bends. The first bend at near zero indicates the sharp edge of the green light at the d line. The second bend at 20 μm indicates the sharp edge of the red light at the C line. Beyond the second bend, there is only blue light at the F line. The region between the bends comprises only red and blue, which appear as a magenta ring of 20-μm width.

The magenta ring is an important metric during focus of the condenser. First, an operator closes the illumination field stop until the edge of the illumination field fits within the vision field. Second, the operator drives the condenser focus toward maximum brightness of the magenta ring. This procedure ensures the best focus of the green d line toward consistent operation of the instrument. A smaller illumination NA might increase visibility of the magenta ring.

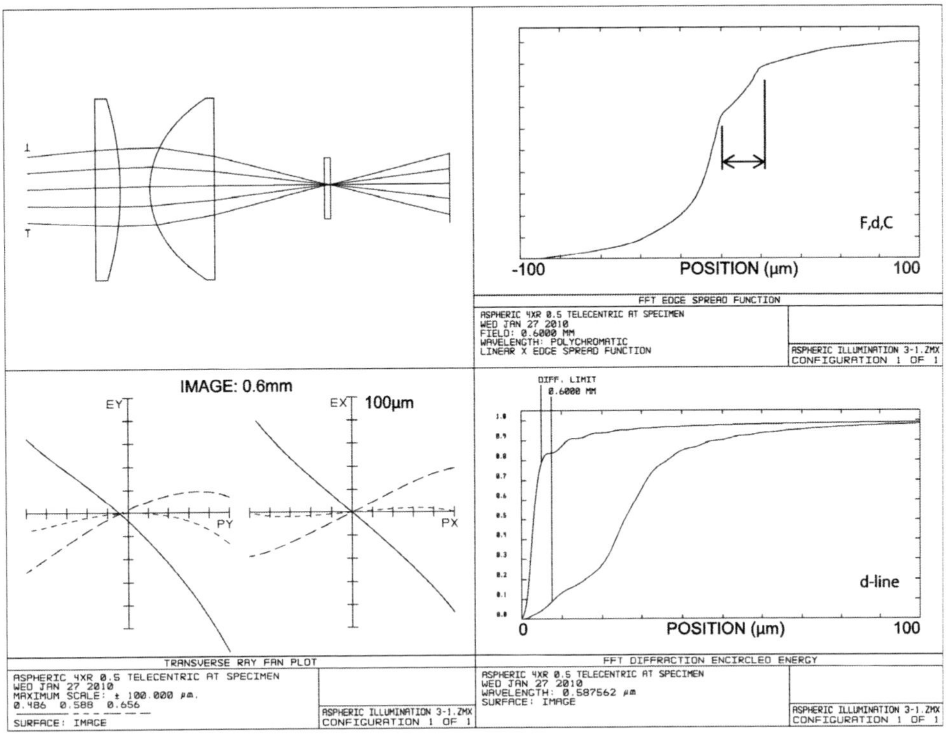

Figure 10.5 Aspheric relay 4XR 0.25. The magenta ring is defined by two bends in the edge spread. Polychromatic edge spread employs 1024 × 1024 sampling. Both lenses are stock production.

10.9 Achromatic Aspheric Plus Doublet Relay

Figure 10.6 displays a lens report for a condenser comprising an aspheric lens and a doublet. The doublet is a custom item, whereas the aspheric lens is a stock product. A small amount of third-order spherical aberration limits the d line to a sharp focus of 10 μm. The axial color limits the polychromatic focus to 40 μm.

10.10 Abbe Condenser

Figure 10.7 displays a lens report for a condenser comprising two plano-convex singlets that are stock items. This combination defines an Abbe condenser, which has numerous formats. A large spherical aberration limits the green d line to a broad focus of 50 μm. The axial color limits the polychromatic focus to 80 μm. Observation of a magenta ring requires an NA of approximately 0.10. Observation of the Abbe sine condition may eliminate the spherical aberration.

10.11 Abbe Aspheric

Figure 10.8 displays a lens report for an aspheric condenser. The axial color limits the polychromatic focus to 50 μm.

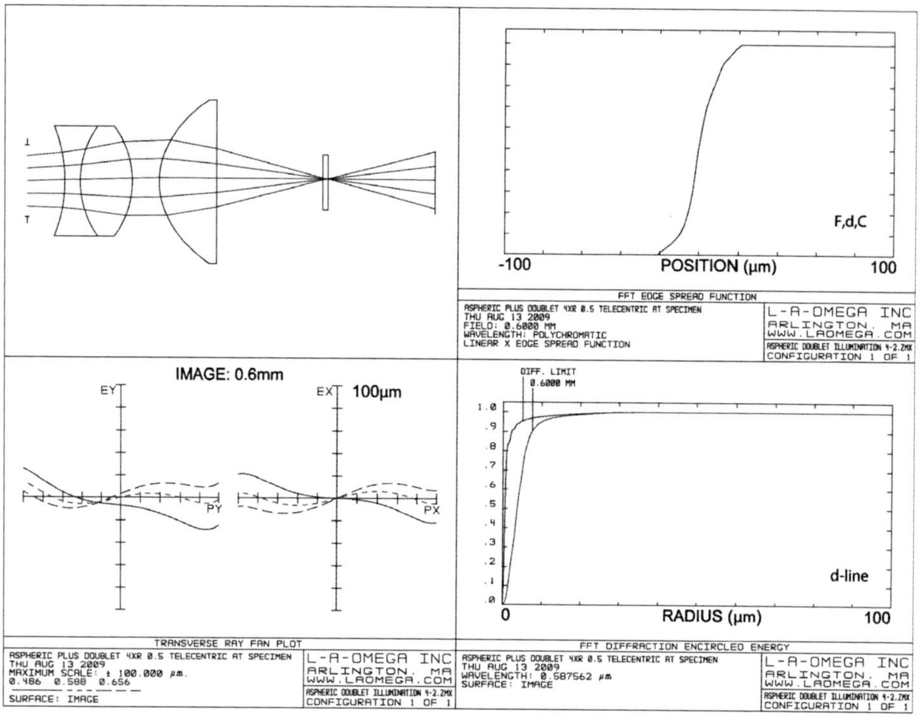

Figure 10.6 Achromatic aspheric relay 4XR 0.25. A custom doublet in combination with a stock aspheric lens.

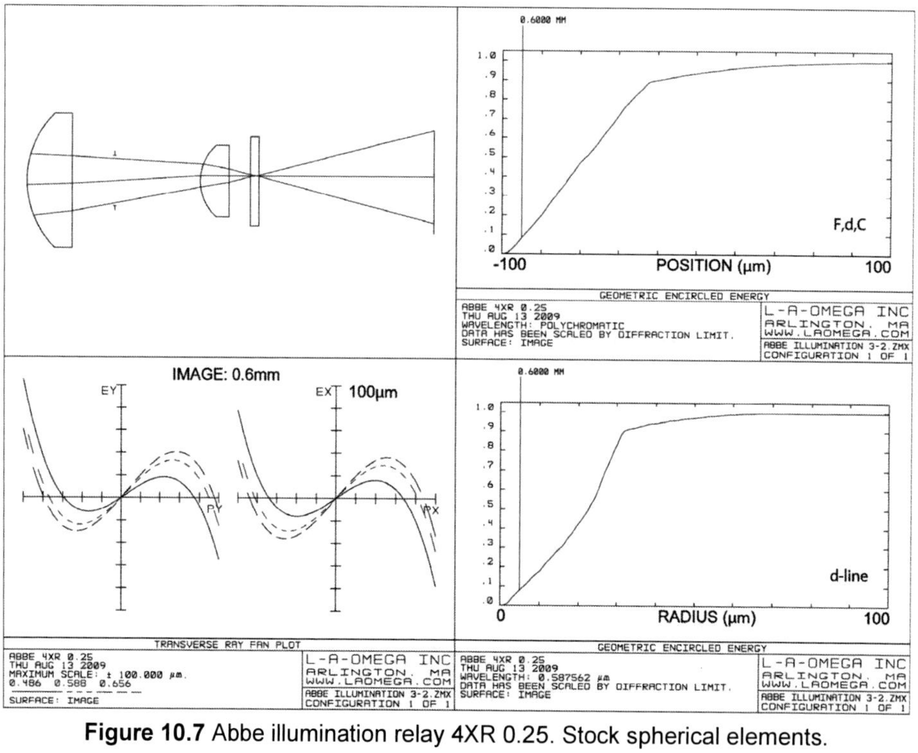

Figure 10.7 Abbe illumination relay 4XR 0.25. Stock spherical elements.

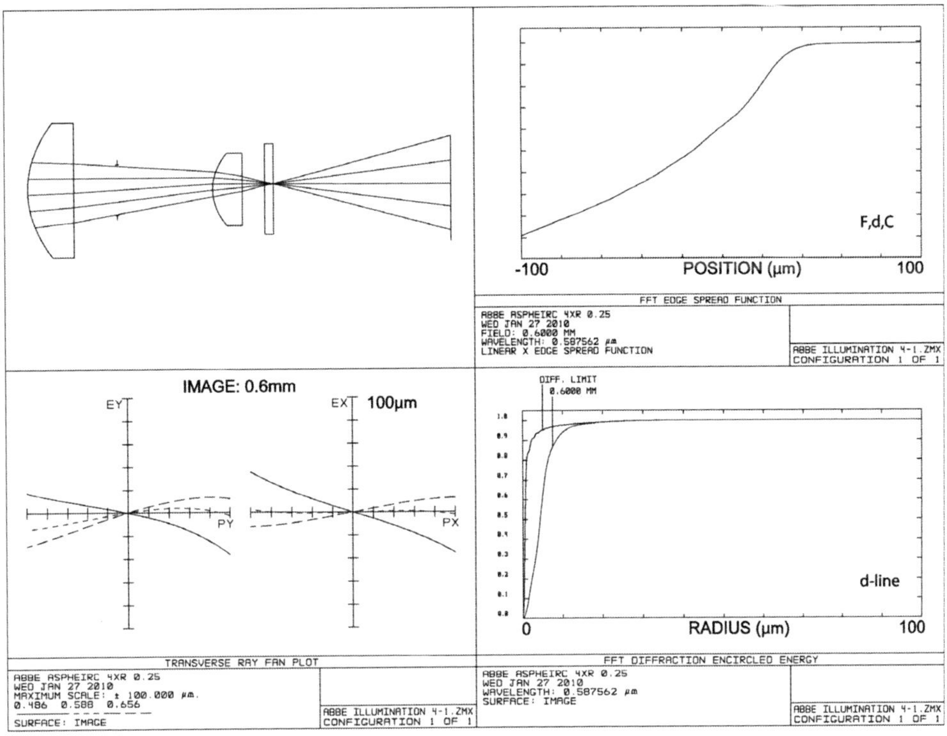

Figure 10.8 Aspheric Abbe relay 4XR 0.25. Custom aspheric elements utlizing B270 glass for molding.

10.12 Total Internal Reflection Fluorescence Illumination

A total internal reflection fluorescence (TIRF) illumination scheme employs total internal reflection for application to fluorescence. The evanescent field of TIR provides a shallow excitation of a liquid on the back side of the cover glass. The depth of the evanescent field is normally less than 100 nm. A small molecule is bound to the back side of the cover. The small molecule is efficiently excited without excitation of the entire depth of the liquid. A single fluorophore can be detected without competition from the intrinsic fluorescence of the bulk of the liquid.

Figure 10.9 displays a TIRF illumination scheme. A laser beam enters the "back aperture" of the lens. A virtual focus is located at the tube focal plane of the TIRF lens. The virtual focus is converted into a collimated beam at the specimen. The real focus occurs inside a calcium fluoride lens element. Purity of calcium fluoride is essential for preventing background radiance from the real focus with large irradiance. Background radiance may be caused by fluorescence and/or nonlinear dipole currents.

The angle of TIR within the cover is

$$\theta_{TIR} = \arcsin\left(\frac{n_L}{n_C}\right), \tag{10.3}$$

where n_L is the refractive index of the liquid and n_C is the refractive index of the cover. The angle of the TIRF beam is described by a thick-lens model, where the principal plane is a sphere within the cover slip. The angle of the TIRF beam in the cover is defined by the height of the beam h_B as follows:

$$\sin \theta_{BC} = \left(\frac{h_B}{n_C f} \right), \tag{10.4}$$

where $n_C f$ is the focal length within the refractive index of the cover. Conversion from glass to silica shortens the immersion focal length. The angle of TIR at water grows from 61 deg in borosilicate to 66 deg in silica.

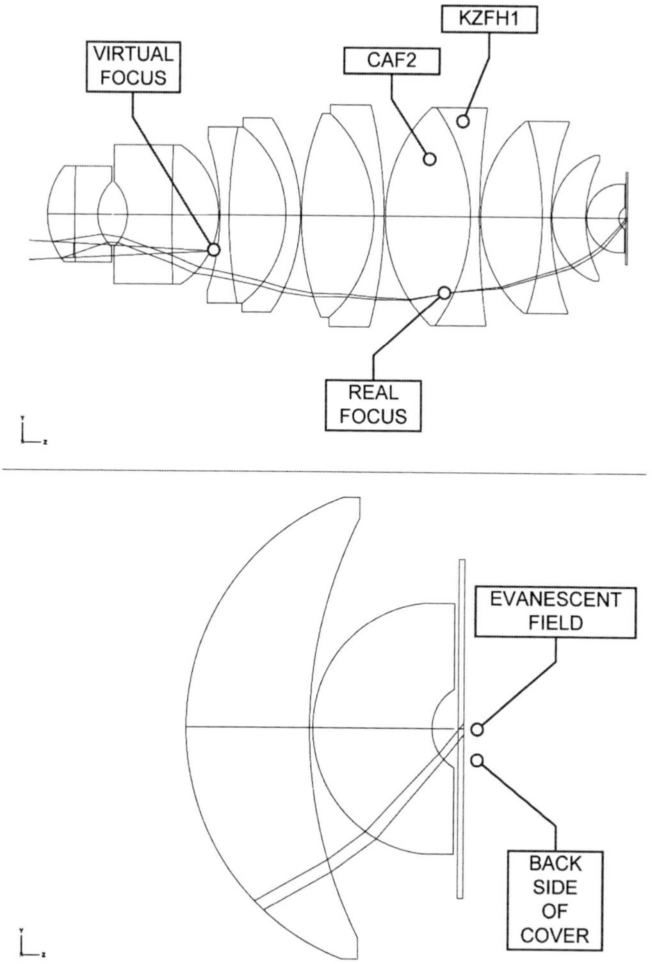

Figure 10.9 TIRF illumination 60X. The TIR of the laser beam provides shallow excitation on the back side of the cover. The angle of TIR is 60.4 deg for D-263 and water. The prescription is the same as the 60X immersion lens of Fig. 12.5.

Chapter 11
Cover Strata

11.1 Importance of Specimen Tolerance

A specimen tolerance budget is essential for consistent performance of an optical instrument. The cover strata should have a specification for nominal thickness and tolerance. Exceeding the specimen tolerance might create enormous spherical aberration. Spherical aberration can reduce contrast without an apparent growth in spot size. Other important issues for the specimen are intrinsic fluorescence, surface-form error, surface defects, and surface texture.

11.2 Perfect 10X for Air

A perfect 10X objective within air is defined by Prescription 11.1 in the Appendix. Equation (2.4) defined the focal length of a 10X objective as 20 mm in combination with a 200-mm tube lens. A typical sensor field is defined by a 12-mm diagonal sensor. Thus, the object field of a 10X objective is 1.2 mm in diameter or 0.6 mm in height.

A virtual object is located at the back focal point. A spherical principal plane defines the lens stop. The object distance from the lens stop is –20 mm in the prescription. The radius of the lens stop equals the back focal length at 20 mm. Rays traveling leftward from the object define paths of perfection without aberration. The rightward-traveling rays after the lens stop should overlay the paths of perfection, unless there is deviation from the cover strata of the virtual object.

The object-space NA is defined within the perfect part of the lens prescription. The lens-stop diameter at 10 mm is defined by the object-space NA. The image-space NA may vary due to aberrations in the cover strata of the image side. A planar lens stop is an acceptable alternative to a spherical lens stop. However, a planar lens-stop diameter requires an inflated diameter for the same object NA. The lens stop and focal length define the *f*/# as 2.0. A spherical lens stop provides a more accurate definition of *f*/#. The distance from the lens stop is variable, as indicated by a box around the thickness of the lens stop.

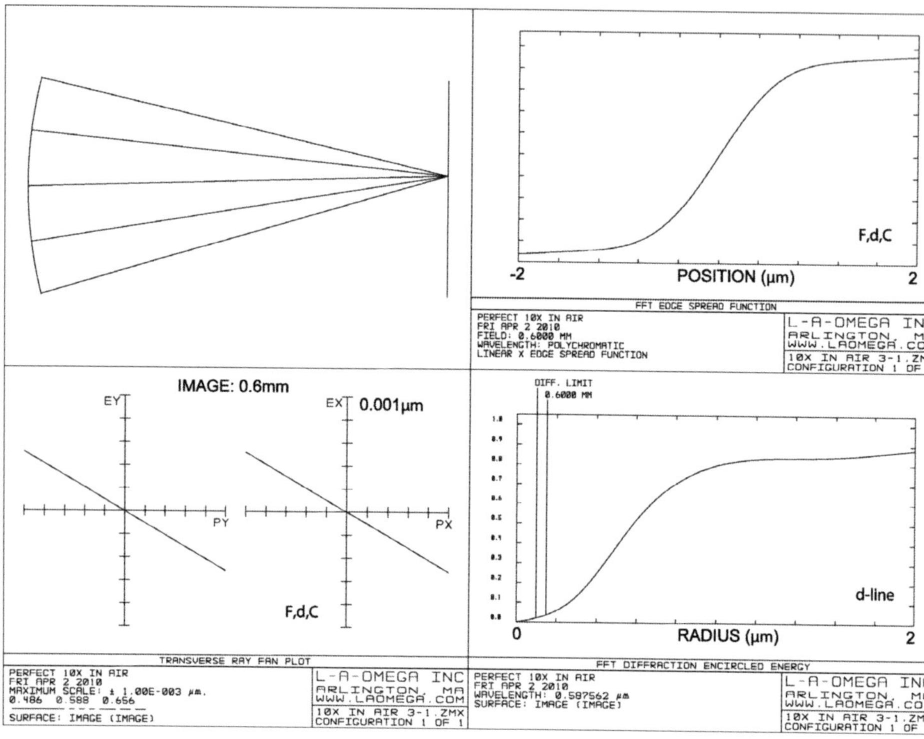

Figure 11.1 Perfect 10X objective in air. Focal length is 20 mm, object NA is 0.25, and the lens stop is 10 mm in diameter. Line-spread is polychromatic, and encircled energy is at the d line. Field height is at 0.6 mm. Axial defocus of 1 nm is intentionally included for visualization of ray-intercept plots.

The corresponding merit function for a perfect 10X objective employs three basic metrics: the ray intercept of a top marginal ray, the ray intercept of a bottom marginal ray, and a Strehl ratio for the d line at the field margin. A process of iteration drives a single variable (lens-stop thickness) toward the maximum Strehl ratio at the field height.

Figure 11.1 displays a lens report for perfect 10X objective within air. A nanometer of axial length beyond perfect focus creates a small amount of defocus. The ray-intercept plots for all three wavelengths are colinear. The amplitude of defocus error is far below the Airy radius at 1.5 μm for the d line. The edge spread is 0.8 μm wide. The flat of the encircled-energy plot ends near 1.5 μm.

11.3 10X Objective with Cover Glass in Place of Air

In Fig. 11.2, a typical cover glass is added to the perfect 10X objective in air. A 0.17-mm thickness of D263M glass represents a No. 1.5 cover slip, which has a manufacturing tolerance of 0.16–0.19 mm. Other standard types of cover glass are cited in Table 11.1. There is a hint of spherical aberration and coma.

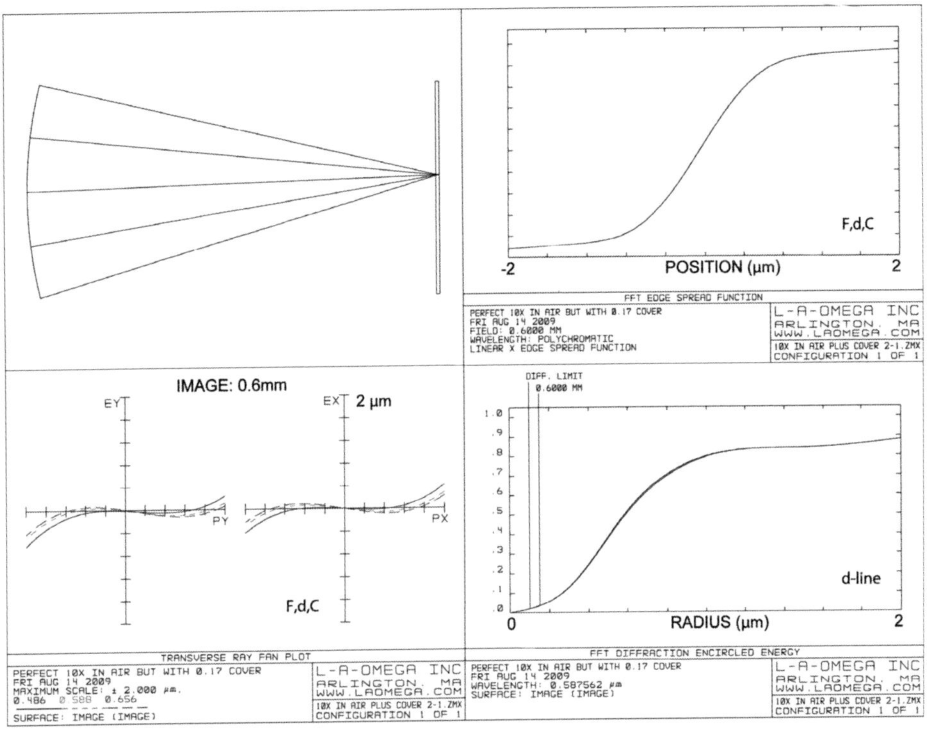

Figure 11.2 Glass cover in image space for a perfect 10X objective in air. Glass cover: D263M at 0.17 mm thick. Focal length is 20 mm, object NA is 0.25, and the lens stop is 10 mm in diameter. Line-spread is polychromatic, and encircled energy is at the d line. Field height is at 0.6 mm.

However, the image is considered diffraction limited. There is a slight departure from the diffraction limit in the encircled-energy curve. A 10X objective at 0.25 NA can operate properly with or without a 0.17-mm cover glass.

A cover glass is manufactured by pulling a ribbon of glass off a molten tin bath. The surface is largely free of digs and scratches. There are no scratches due to pulling of impurities during polish. However, there are significant tolerances of thickness, which become more important as the objective NA becomes larger. Table 11.1 displays a list of typical cover-glass tolerances.

11.4 10X Objective with Microscope Slide in Place of Air

In Fig. 11.3, a typical microscope slide is added to the perfect 10X objective in air. A 1.0-mm thickness of D263M represents a typical microscope slide. There is enough spherical aberration for a small departure from the diffraction limit of the d line. There is an even smaller amount of axial color as indicated by the dissimilar slopes at the origin of the ray intercepts. The Strehl ratio of the margin is 96% for the d line and 91% for the polychromatic combination of the F, d, and C lines. A 10X objective at 0.25 NA can operate properly with a 1.0-mm glass slide.

Table 11.1 Cover glass tolerances.

Number	Nominal	Minimum	Maximum	Range	Typical area
	Dimensions (mm)				
0	0.10	0.085	0.115	0.030	15 × 15 to 24 × 60
1	0.15	0.130	0.160	0.030	15 × 15 to 24 × 60
1.5	0.17	0.160	0.190	0.030	15Ø, 15 × 15, 24 × 50
2	0.21	0.190	0.230	0.040	
3	0.30	0.280	0.320	0.040	
4	0.37	0.320	0.420	0.040	24 × 50
5	0.55	0.500	0.600	0.100	
Slide	1.00	0.960	1.060	0.100	25 × 75
Slide	1.20	1.100	1.200	0.100	25 × 75

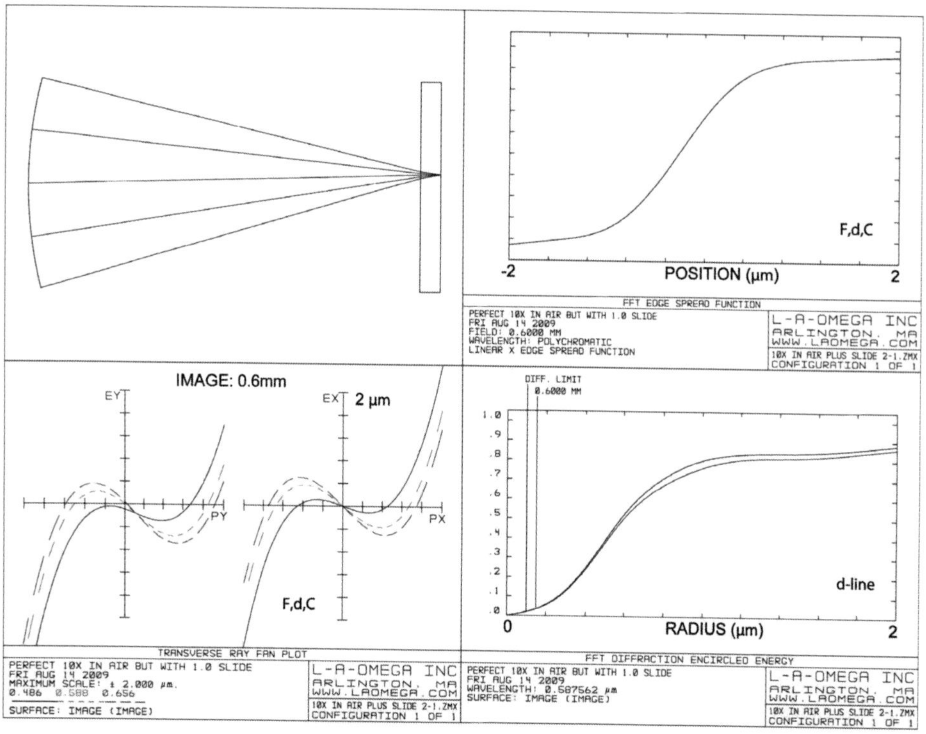

Figure 11.3 Glass slide in image space for a perfect 10X objective. Glass slide: D263M at 1.0 mm thick. Focal length is 20 mm, object NA is 0.25, and the lens stop is 10 mm in diameter. Line spread is polychromatic, and encircled energy is at the d-line. Field height is at 0.6 mm.

11.5 40X Objective with Silica Cover in Place of Glass

Prescription 11.4 (see Appendix) defines a cover glass within a perfect object space. The correct cover glass is 0.17 mm thick. The correct cover material within object space is D263M. The effective focal length of the 40X objective is 5.0. The optical distance [Eq. (3.24)] from the lens stop to the object is also 5.0

mm. The spatial distance is > 5.0 mm due to immersion with the cover glass. On the image side, silica replaces D263M as the cover material. The material in image space may be glass, silica, water, or air. Silica is used in this example for simplicity. At higher NAs with immersion oil, the choice of image material becomes important as total internal reflection becomes possible.

Figure 11.4 displays a perfect 40X objective with silica in place of D263M on the image side. The object-space NA is 0.75. The image-space NA is 0.74. There are small amounts of axial color and third-order spherical aberration. There is a large fifth-order spherical aberration at the margin. However, the encircled-energy plot indicates diffraction-limited performance. The Strehl ratio of the margin is 100% for the d line and 98% for the polychromatic combination of the F, d, and C lines. Thus, a 40X objective at 0.75 NA can operate properly with a silica cover in place of a glass cover. However, the NA cannot increase with inclusion of a larger spherical aberration.

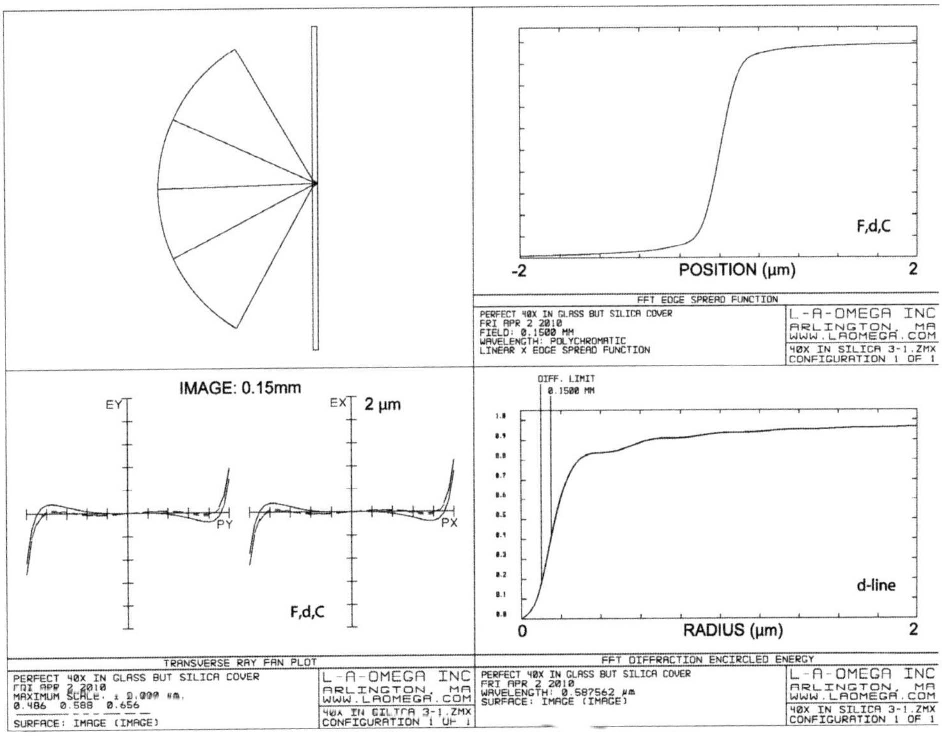

Figure 11.4 Silica specimen cover replacing a glass cover in image space for a perfect 40X objective. Silica is used in place of D263M for the cover material at 0.17 mm thick. Focal length is 5 mm, object NA is 0.75, and the lens stop is 8.6 mm in diameter. Line-spread is polychromatic, and encircled energy is at the d line. Field height is at 0.15 mm.

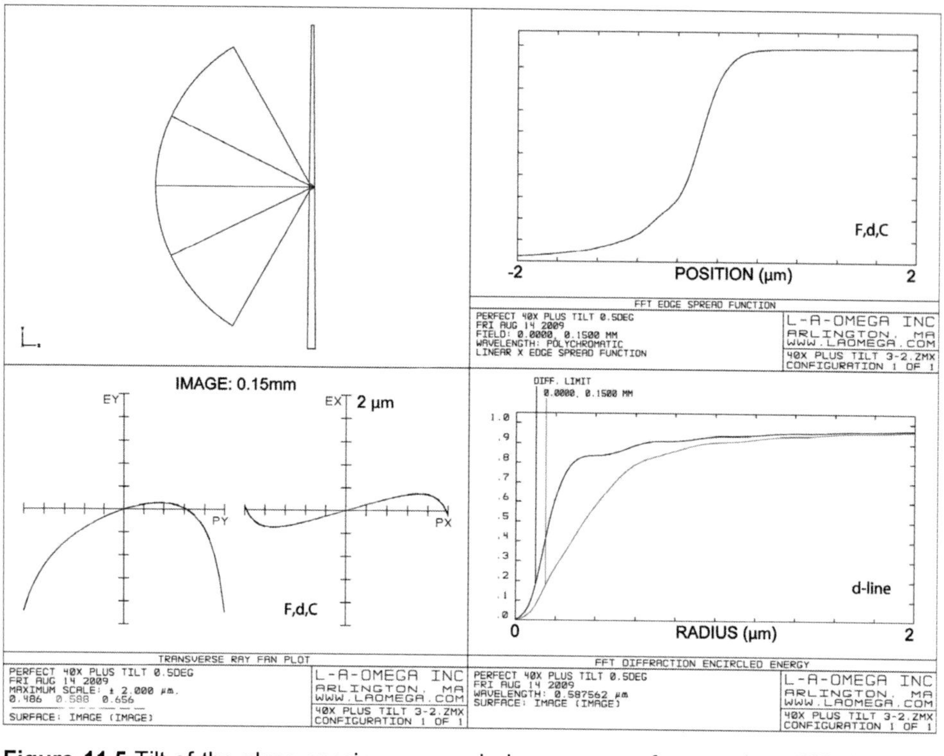

Figure 11.5 Tilt of the glass specimen cover in image space for a perfect 40X objective. Cover is at a tilt of 0.5 deg (9 mrad). Cover material is D263M at 0.17 mm thick. Focal length is 5 mm, object NA is 0.75, and the lens stop is 8.6 mm in diameter. Line-spread is polychromatic, and encircled energy is at the d line. Field height is at 0.15 mm.

11.6 40X Objective with Tilted Cover Glass

Figure 11.5 displays a perfect 40X objective with a specimen tilt of 0.5 deg (9 mrad). Three object points are required for optimization: 0.0, +0.15, and –0.15 mm. An image tilt of –1.0 deg in silica is required as compensation for the +0.5-deg tilt of the cover within image space. The object-space NA is still 0.75. There is a large amount of coma, which blurs the left edge of the line spread. There are much smaller amounts of astigmatism and fifth-order spherical aberration. The encircled-energy plot indicates a large departure from the diffraction limit. The Strehl ratios for the d line are 78% at 0.15 mm, 64% at +0.15 mm, and 85% at –0.15 mm. A 40X objective at 0.75 NA is significantly compromised by coma due to a 0.5-deg tilt of the cover glass.

There is also significant tilt of the specimen plane at the sensor. The tilt of the specimen plane within the cover glass is –0.5 deg with respect to the optical axis. The equivalent tilt of the specimen plane within air is approximately 0.3 deg. After 40X magnification, the image of the specimen plane at the sensor is tilted by 12 deg. The amplified effects of specimen tilt demand careful consideration at any magnification.

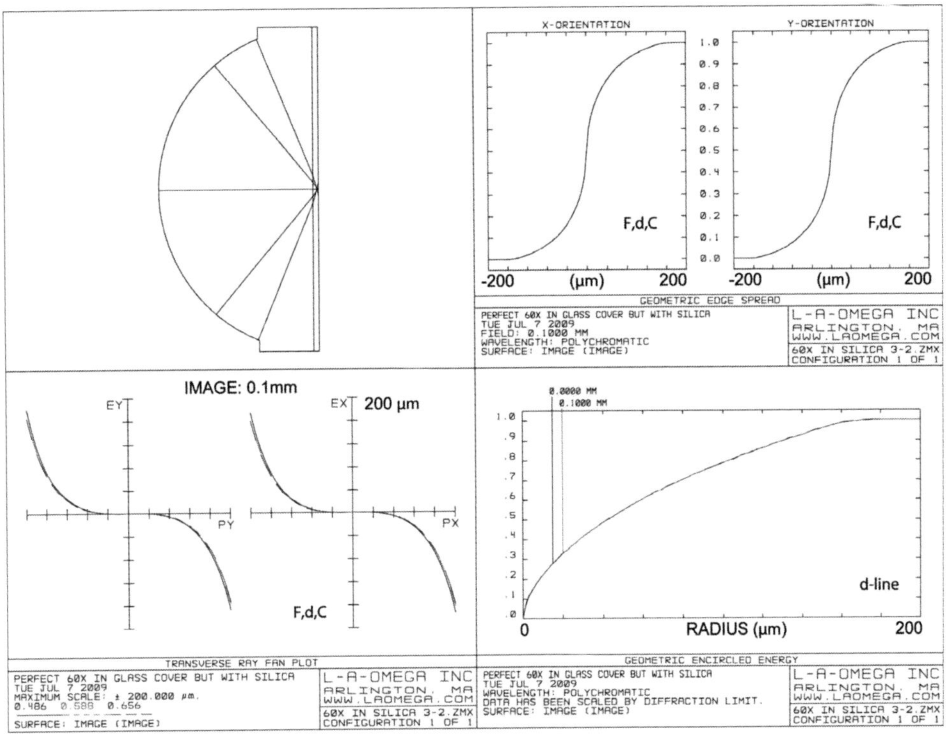

Figure 11.6 Silica specimen cover replacing a glass cover in image space for a perfect 60X objective. Silica is used in place of D263M for the cover material at 0.17 mm thick. There is immersion oil between the lens stop and cover. Focal length is 3.3 mm, object NA is 1.40, and the lens stop is 9.3 mm in diameter. Line-spread is polychromatic, and encircled energy is at the d line. Field height is at 0.10 mm.

11.7 60X Objective with Silica Cover in Place of Glass

Figure 11.6 displays a perfect 60X objective lens with silica in place of D263M as the cover material. Immersion oil fills the gap between the lens stop and the cover. The focal length is 3.3 mm. The immersion focal length is 5.0 mm. The radius of curvature of the lens stop is 5.0 mm. The object-space NA is 1.40. The image-space NA is 1.35. The diameter of the lens stop is 9.3 mm. There is an enormous amount of third-order spherical aberration. The Strehl ratio of the margin is 2% for the d line. A 60X objective at 1.40 NA cannot operate properly with a silica cover in place of a glass cover.

11.8 Strehl Ratio versus Optical Path Length

Due to additional cover strata, the Strehl ratio is very dependent on marginal NA. Figure 11.7 displays the Strehl ratio versus marginal NA for several common errors. The plots are primarily identified by optical path lengths as 50λ, 200λ, and 2000λ. These path lengths represent three common thickness errors: a cover glass at 30 μm of thickness tolerance, a microscope slide at 120 μm of tolerance, and a microscope slide at 1.2 mm in thickness. The 50λ path indicates a maximum NA of 0.50 before a cover tolerance degrades the spot; the 200λ path indicates a maximum NA of 0.35 before a slide tolerance degrades the spot; the 1200λ path indicates a maximum NA of 0.20 before a slide degrades the spot.

Consideration of specimen tolerances is critical for consistent performance of an instrument. Application of a binder adds significant amounts of unexpected optical path length when using a glass cover. The tolerance on the binder thickness might exceed the nominal thickness of the cover. A large amount of binder might enable tilt of the cover. A credible tolerance for a specimen should be defined before an optical instrument is designed. Consistent performance is a mark of quality in optical design and manufacture.

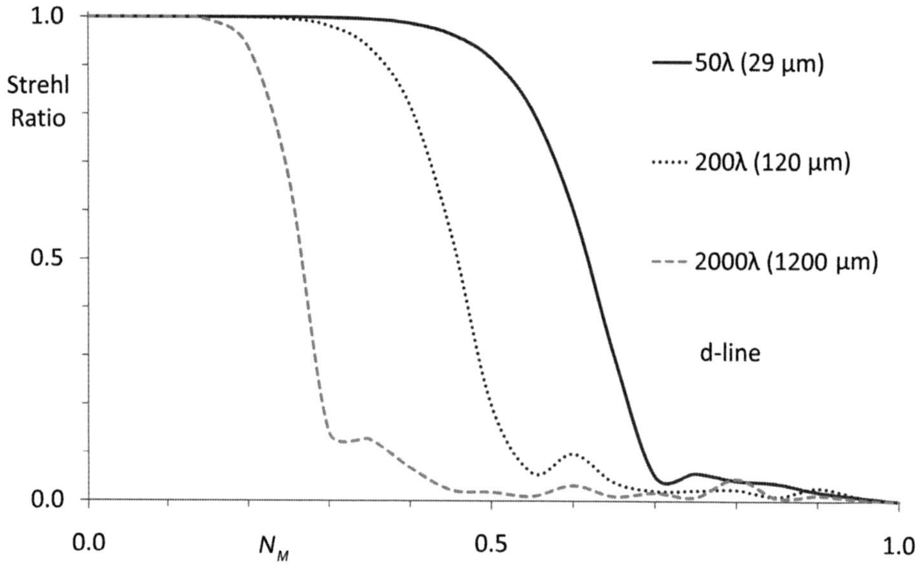

Figure 11.7 Strehl ratio versus optical path length of an additional cover. Optical path lengths are cited in wavelengths. Spatial thickness of D263M is cited in parentheses. 50λ (29 μm) indicates a tolerance for a cover glass. 200λ (120 μm) indicates a tolerance for a microscope slide. 2000λ (1200 μm) indicates a thickness for a microscope slide.

Chapter 12
Objective Lenses

12.1 Formats

There are numerous types of objectives; several are listed below:

1. An aplanat[i] is free from spherical aberration and coma.
2. A plan objective has a flat field: there is no Petzval curvature.
3. An achromat is corrected at two wavelengths.
4. An apochromat is corrected at three wavelengths.
5. A semiapochromat is nearly corrected at three wavelengths.

The original fluorite objectives corrected for color with fluorite glass, but lanthanum glass eventually replaced fluorite in most applications. A modern fluor objective contains LaK while excluding many schwer flints (SF). A fluor objective transmits well in the UV; consequently, a fluor objective may also imply application to fluorescence.

The magnifications of the following objectives are based on a 200-mm tube lens, which is a standard tube lens for Nikon microscopes. An Olympus scope employs a 180-mm focal length as its tube lens.

12.2 Aplanatic Surface

An aplanatic surface is free of both spherical aberration and coma. There are three configurations of an aplanatic surface. An aplanatic surface of the flat kind has little practical value: the object and image are located at the surface. An aplanatic surface of the concentric kind is very intuitive: the object and image are concentric with the surface. There is no refraction in the second configuration. An aplanatic surface of the scaled-divergence kind is not obvious: it is defined by the following mathematical relation, where the divergence is scaled by the refractive index:

$$\frac{\mathrm{NA}_I}{n_I} = \frac{\mathrm{NA}_O}{n_O}, \tag{12.1}$$

[i] Aplanat is derived from the Greek *a–plan-etes*. *a-*, meaning not; *plan*, meaning moving; and *etes*, meaning star. However, an aplanat does not transform a star into a planet during an exposure.

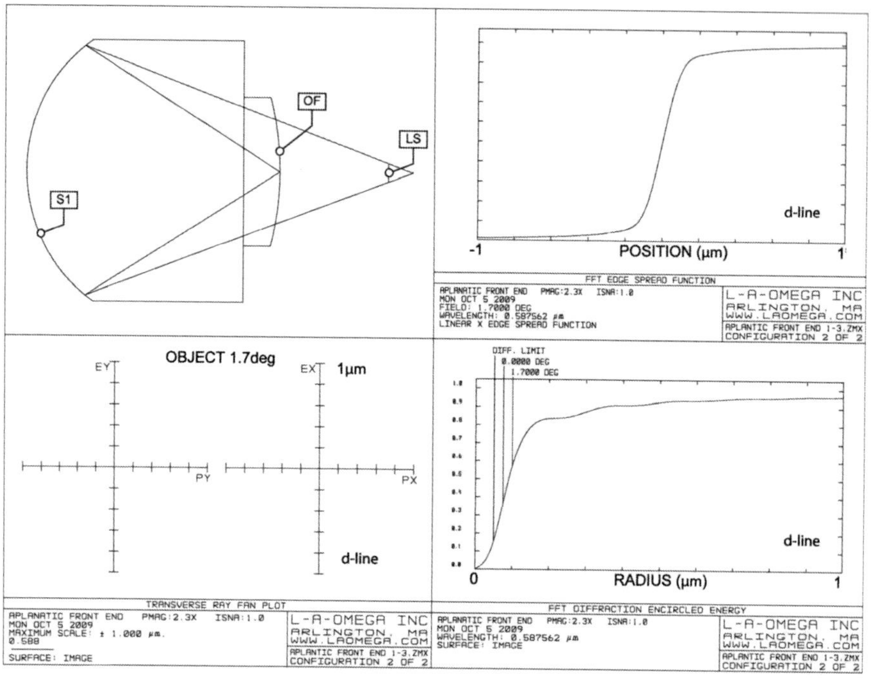

Figure 12.1 Aplanatic surface of the scaled-divergence kind. The NA is scaled by the refractive index between the object and the image. The lens stop is proximal to the image.

where NA_I is the NA of the image, n_I is the refractive index of the image, NA_O is the NA of the object, and n_O is the refractive index of the object.

Figure 12.1 displays a single aplanatic surface of the scaled-divergence kind. This eliminates the spherical aberration. The divergence within the glass borosilicate crown (BK7) is roughly 1.5 times that of air. Proper placement of the lens stop near the image conjugate also eliminates coma. The image field is curved. The aplanatic surface is free of both spherical aberration and coma.

In Fig. 12.2, the addition of a meniscus defines two more aplanatic surfaces. The first and third surfaces S1 and S3 are aplanatic due to scaling of divergence by the refractive index. The second surface S2 is aplanatic due to concentric image conjugates of the surface. Once again, proper location of the lens stop is required for elimination of coma, and the object field is curved.

12.3 10X Plan Achromat

Figure 12.3 displays a 10X plan achromat.[21] The focal length is 20 mm. The paraxial NA is 0.25. In the ray-intercept plot, correction at two wavelengths (588 and 656 nm) indicates an achromat, and lack of linear slope indicates plan. The encircled-energy plot indicates nearly diffraction-limited performance at 0.6 mm and slightly aberration-limited performance at 0.6 mm. The flat of the encircled-energy plot indicates an Airy radius of 1.2–1.4 μm. A 0.25 NA indicates an Airy radius of 1.4 μm at 588 nm. The edge spread displays a high-contrast transition over 1.0 μm. The SCE is 1.9% in air and 0.9% in water.

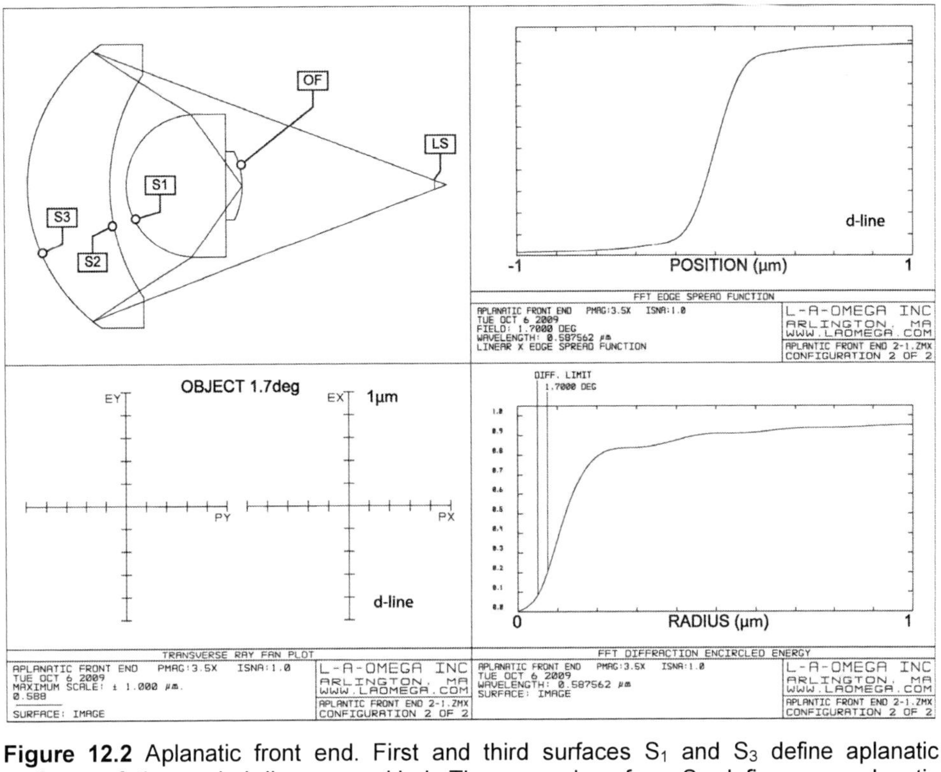

Figure 12.2 Aplanatic front end. First and third surfaces S_1 and S_3 define aplanatic surfaces of the scaled-divergence kind. The second surface S_2 defines an aplanatic surface of the concentric kind.

There are two distinct lens groups. The first group resembles a single Gauss: a biconvex singlet and meniscus doublet. The meniscus lens is shaped for management of spherical aberration, coma, lateral color, and Petzval curvature. The second group is a doublet: it manages the flange focal distance and the residual aberrations of the first group.

The lens employs super-schwer crown (SSK), lanthanum schwer flint (LaSF), and schwer flint (SF). The SSK glass has a refractive index of approximately 1.6; the LaSF and SF have a large refractive index at 1.8. Performance in the blue is compromised for two reasons: the partial dispersion and absorption in the blue spectrum. Both limitations originate in the SFs.

This lens functions extremely well for human vision, although its performance in the blue compromises electronic vision.

12.4 40X Fluor

Figure 12.4 displays a 40X fluor.[22] The focal length is 20 mm. The paraxial NA is 0.75. The original lens had a focal length of 1 mm and a 0.17-mm-thick cover glass. The lens was scaled to 5 mm in focal length while maintaining the cover glass at 0.17 mm thick. The entrance pupil at 7.5 mm was derived from a 0.75 NA. The flat of the encircled-energy plot indicates an Airy radius of 0.4–0.5 µm.

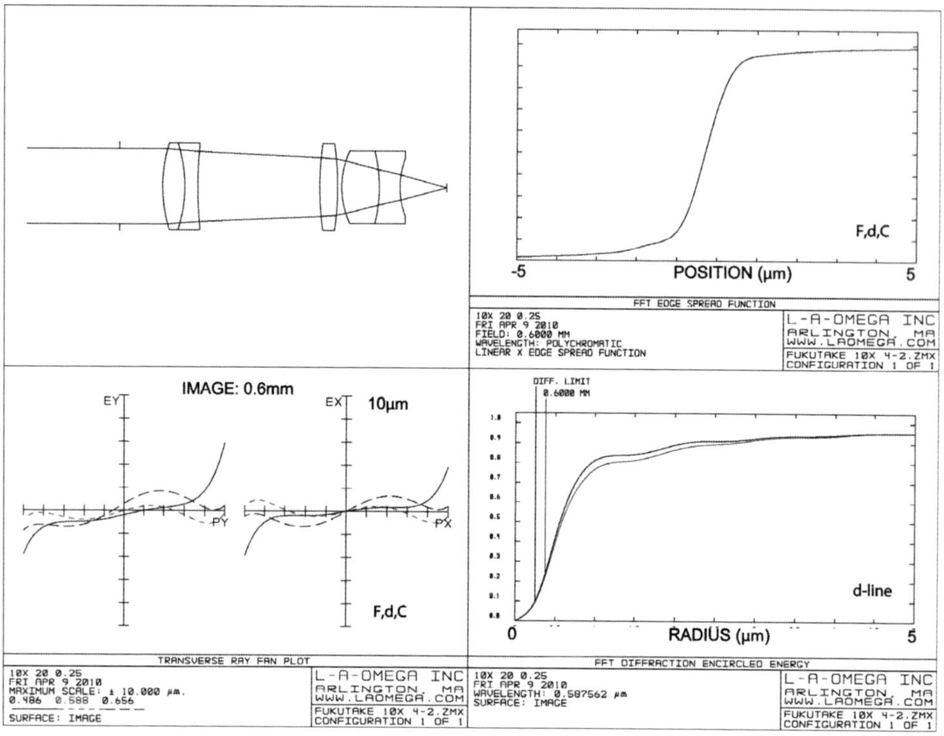

Figure 12.3 10X plan achromat at 0.25 NA. NA_P = 0.25 is derived from a 5.0-mm entrance pupil and a 20-mm EFL. Prescription by Fukutake.[21]

An NA of 0.75 indicates an Airy radius of 0.5 µm at 588 nm. The Gaussian depth is 1.9 µm. The edge spread displays a high-contrast transition over 0.3 µm. The SCE is 8.7% in water.

The ray-intercept plot indicates several key features. The lens is focused for the d line. There is a small amount of spherical aberration at the margin, which is likely clipped in application. The d and F lines are effectively corrected with each other. The lens is an achromat, even though the patent specifies a semiapochromat. The lens is fairly corrected for the C, d, and F lines, because they all lie within an axial range of 3 µm. However, the g line (436 nm) displays significant axial color at 9 µm. Perhaps in the current lens, semiapochromat indicates correction for three wavelengths within a tolerance, but not a fourth.

There are three distinct groups of the lens, and they each serve a primary purpose. Optimization of the prescription may require a small deviation from the group's primary role.

The first group resembles an aplanatic front end. The first surface is an aplanatic surface of the concentric kind. The second surface is an aplanatic surface of the scaled-divergence kind. This nearly hemispheric lens employs a high-index lanthanum flint. The third surface is an aplanatic surface of the concentric kind, and the fifth surface is an aplanatic surface of the scaled-divergence kind. The third through fifth surfaces define an aplanatic doublet.

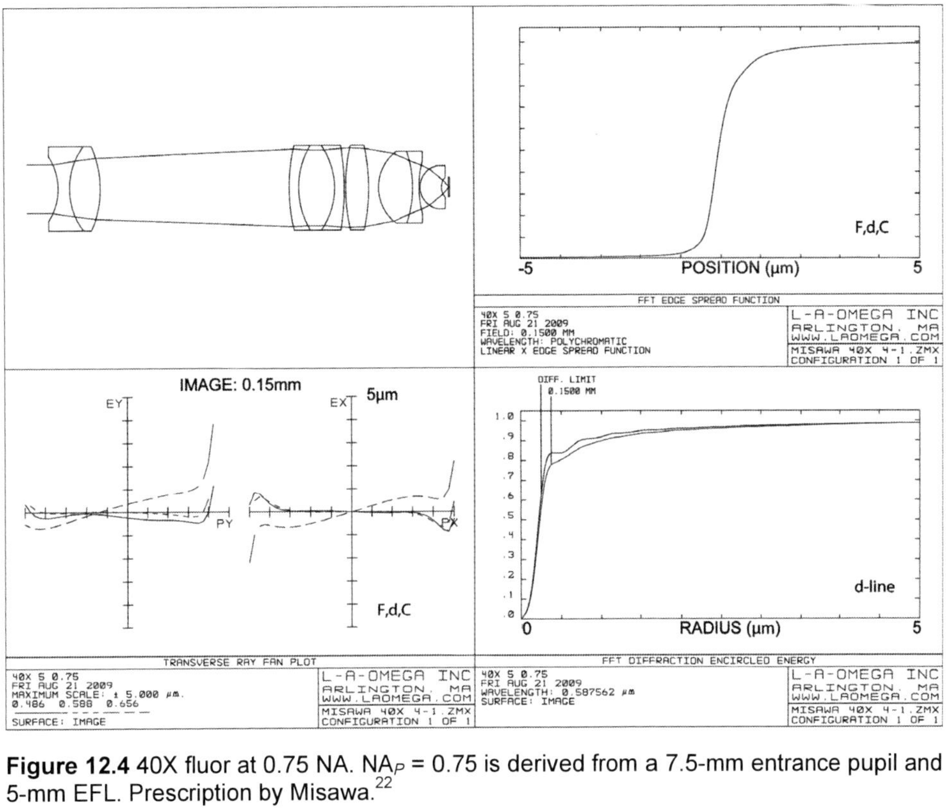

Figure 12.4 40X fluor at 0.75 NA. NA_P = 0.75 is derived from a 7.5-mm entrance pupil and 5-mm EFL. Prescription by Misawa.[22]

The second group resembles a single Gauss, where the doublet is a triplet. The positive elements are all long-spectrum fluorite krone glass: FK5 and FK56 provide a low index and low dispersion. The low index limits the NA of the objective, while the low dispersion reduces lateral color.

The third group defines a negative element for extension of track length beyond the focal length. The distance from lens stop to object should be 60 mm, while the effective focal length is 5 mm. This configuration requires a strong negative element at the lens stop. It is somewhat aplanatic with respect to the lens stop.

12.5 60X Immersion TIRF

Figure 12.5 displays a 60X immersion lens.[23] It can be employed for total internal reflection fluorescence (TIRF). See Sec. 10.12 for the illumination path of TIRF.

The paraxial NA is 1.4, the marginal NA is 1.2, and the central NA is approximately 1.0. A hemispheric surface defines the last air-to-glass interface of the lens. The next concave surface holds a drop of oil. The oil has a refractive index of 1.518 and an Abbe number of 58.9. The cover material is likely D-263, which has a refractive index of 1.5255 and Abbe number 55. The cover is 0.17

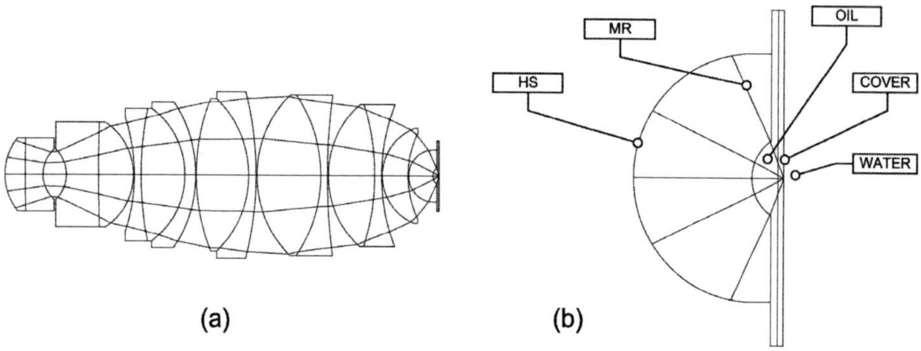

Figure 12.5 60X immersion lens. Rays indicate an NA_P as 1.4. NA_M is 1.2 and NA_C is approximately 1.0. Part (a) displays 8 lens groups and 15 lens elements. Part (b) displays group 8 comprising: a nearly hemispheric surface (HS), an oil immersion, a cover glass, and water as the object medium. Prescription by Yamaguchi.[23]

mm thick. The working distance between the lens and cover is 0.15 mm. The object is located in water on the back side of the cover. The object may also be located within a binder.

The 60X immersion lens is suitable for TIRF. The marginal ray (MR) at 70.2 deg is beyond the critical angle of 60.7 deg at the cover-to-water interface. This enables an evanescent field with a < 100-nm depth into the water. Small molecules may bond to the back surface of the cover. The evanescent field excites the surface-bound molecules without excitation of the bulk liquid.

The 60X TIRF lens employs a sequence of four aplanats at the object. The first aplanat comprises a doublet of oil and lanthanum schwer flint (LaSF35). The convex surface of oil is an aplanatic surface of the concentric kind. The large index of LaSF35 at 2.02 reduces the exiting NA and reduces the spherical aberration. The convex surface of the LaSF35 is an aplanatic surface of the scaled-divergence kind. Three of the aplanats employ a combination of a long-spectrum krone CaF_2 and a short flint KzFH1. The combination of calcium fluoride and lead borate is a common method for color correction in the blue. A short flint from 2009 may be comprised of neodymium and silicate. These four aplanats gradually bend the marginal rays inward towards the optical axis.

At the back of the lens, there are two meniscus doublets. The opposing concave surfaces create a powerful negative lens. Their proper alignment is very difficult to attain. However, their negative power is essential for maintenance of flange focal length and compensation for aberrations of positive aplanats.

Figure 12.6 displays a report for the 60X TIRF lens at 1.4 NA. The ray-intercept plot displays a large spherical aberration of 18 μm at the margin for the d line. The marginal ray angle at 70.2 deg within the cover specifies a marginal NA at 1.23, which is far below the NA. The encircled-energy plot indicates a central peak within 0.4 μm and a broad base beyond 5 μm. Also within the encircled-energy plot, the flat of the 0.1-mm field occurs at 0.4 μm, while the flat of the diffraction limit is at 0.25 μm. An Airy diameter of 0.8 μm at 588 nm

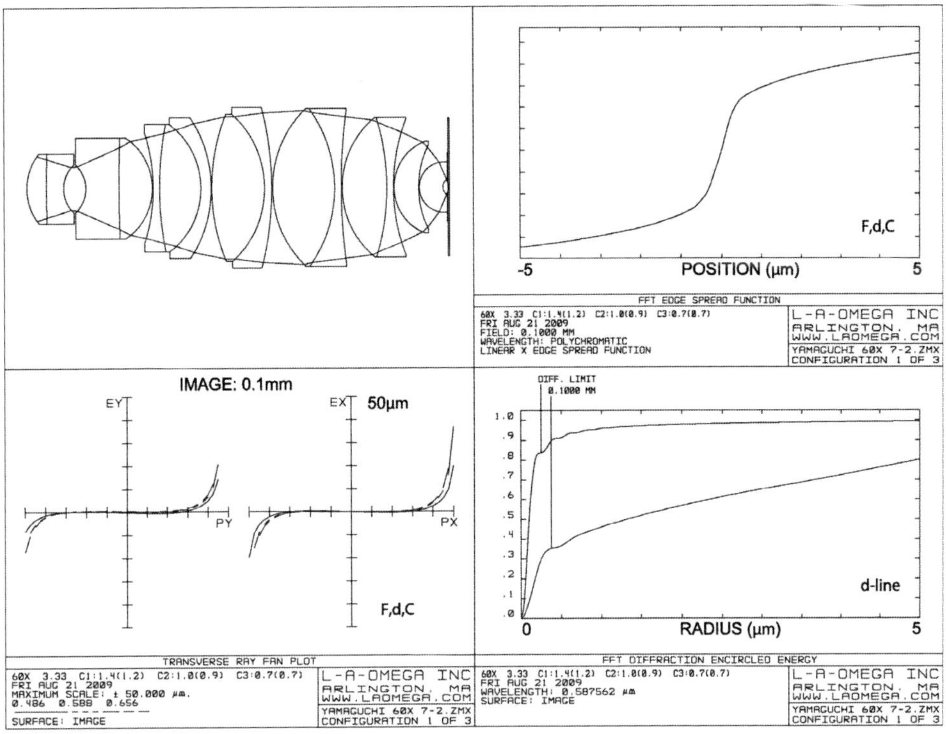

Figure 12.6 60X immersion lens at 1.4 NA. NA_P = 1.4 is derived from a 9.6-mm entrance pupil and a 3.37-mm EFL. NA_M = 1.23. NA_C = 0.90. Edge spread is at the d line and 0.1-mm field. Prescription by Yamaguchi.[23]

indicates a central NA of 0.90. The edge spread displays a sharp edge with poor contrast. The edge rises quickly over 0.2 µm, while the tails extend beyond 5 µm.

Figure 12.7 displays a report for the 60X TIRF lens at 1.0 NA. The ray-intercept plot displays a small spherical aberration of 2 µm at the margin for the d line. The axial color indicates < 3 µm of axial shift. The marginal ray angle at 40.7 deg within the cover specifies a marginal NA at 0.93, which is close to the paraxial NA. The Gaussian depth of focus is 0.43 µm at 588 nm. The encircled-energy plot indicates a central peak within 0.3 µm without a broad base. Also within the encircled-energy plot, the flat of the 0.1-mm field is nonexistant, while the flat of the diffraction limit is also at 0.35 mm. An Airy diameter of 0.8 µm at 588 nm indicates a central NA of 0.9. The edge spread displays a sharp edge with excellent contrast. The edge rises quickly over 0.2 µm with very little tail. Figure 12.8 displays the SCE by spot radius of the 60X immersion lens. The SCE plot is an encircled plot, which is scaled by spherical collection efficiency.

The SCE for 0.7 NA is quite different from the SCEs of 1.0 NA and 1.4 NA. The SCE at 0.7 NA indicates a different central peak and a much different magnitude. The flat indicates an Airy radius at 0.45 µm. The maximum magnitude is 0.07. The area of the central peak is nearly twice that of 1.0 NA.

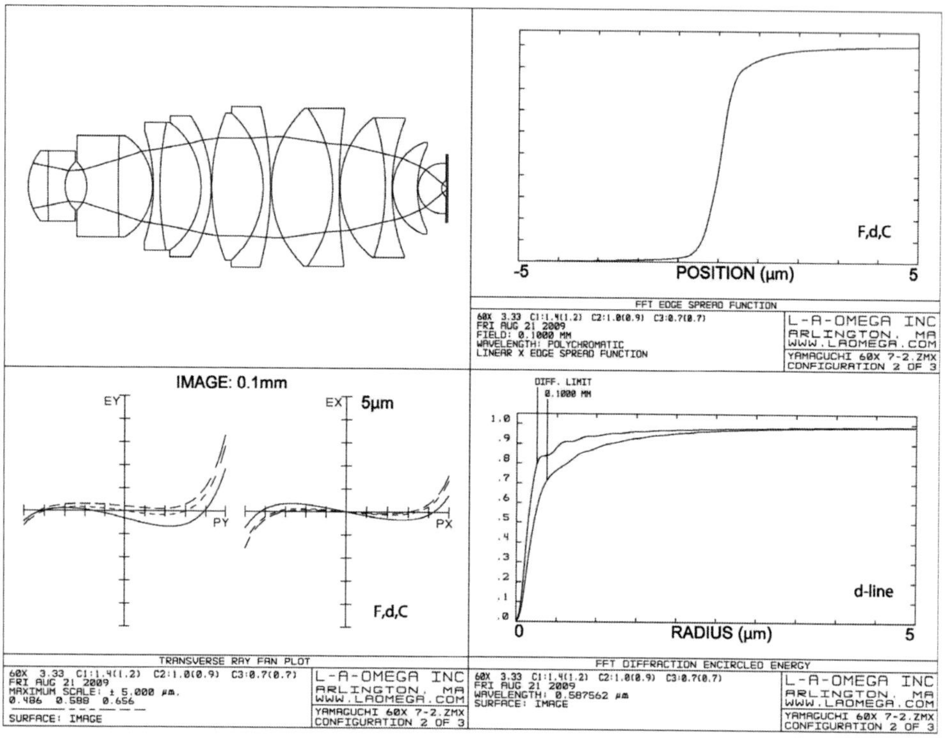

Figure 12.7 60X immersion lens at 1.0 NA. NA_P = 1.0 is derived from a 6.7-mm entrance pupil and a 3.37-mm EFL at the d line. NA_M = 0.93. NA_C = 0.9. Prescription by Yamaguchi.[23]

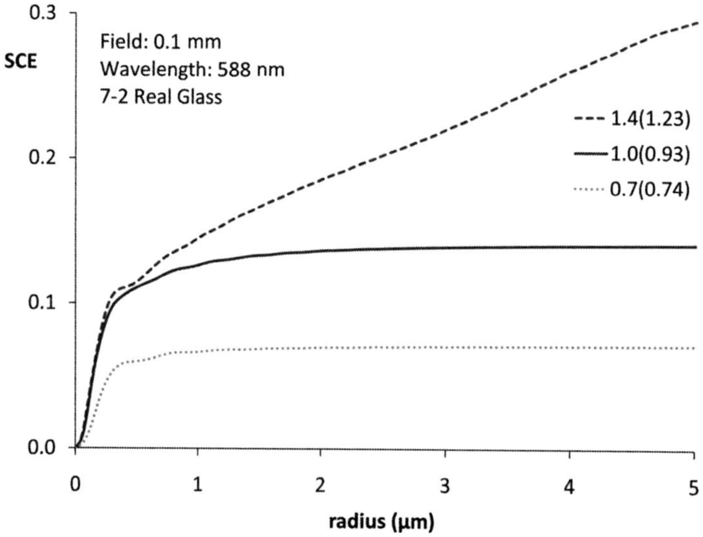

Figure 12.8 SCE of a 60X immersion lens. The SCE is plotted versus radial position within the point spread. The SCE plot is an encircled-energy plot from ZEMAX which is scaled by the SCE. The encircled-energy plot occurs in cover glass (n = 1.52). Spherical collection occurs in water (n = 1.33). Paraxial NAs at 1.4, 1.0, and 0.7, respectively. Marginal NAs at 1.2, 0.9, and 0.7, respectively.

The spherical collection is nearly half that of 1.0 NA. Consequently, the irradiance of the central peak at 0.7 NA is nearly one-quarter that of 1.0 NA. Increasing the NA greatly increases the irradiance of the point spread. However, this trend is not always true. At some point the increased NA might direct more light into a broad skirt without increasing the peak irradiance.

The SCEs at NAs of 1.0 and 1.4 are similar at small radii, but much different at large radii. The SCE for 1.4 NA is slightly larger that of 1.0 NA from 0–0.3 μm. This indicates a common central peak in the point spread. However, the SCE for 1.0 NA remains flat beyond 1 μm, while the SCE for 1.4 NA grows steadily until 4 μm. This indicates a broad skirt in the point spread for 1.4 NA, which is absent in the point spread for 1.0 NA. The maximum SCE for 1.4 NA is 0.31, and the maximum SCE for 1.0 NA is 0.14. The similarity at small radii is due to shared central NA where the lens is diffraction limited. The additional annulus of the 1.4 NA does not tighten the spot because the wavefront of the annulus has spherical aberration.

In combination with a tube lens at 200-mm focal length, the 60X objective creates a 21-μm radius for the central peak at both 1.4 and 1.0 NA. The skirt of 1.0 NA extends to a 60-μm radius, while the skirt of the 1.4 NA extends to a 240-μm radius. At 8 μm per pixel, the skirt of the 1.4 NA extends to a radius of 30 pixels. The central peak of both extends to a radius of 3 pixels. The central peak is 5 pixels wide, the skirt for the 1.0 NA is 15 pixels wide, and the skirt for the 1.4 NA is 60 pixels wide.

Half of the light collected by the 1.4 NA is directed into the background of the image. The shot noise of the skirt is nearly equal to that of the central peak. The skirt noise of a single point spread is spread over many pixels. It might not be above the read noise of the sensor. It might not be detected in a single point spread. However, the skirt noise of multiple point spreads can be significant. The contrast of a central peak with background can be greatly reduced by the summation of multiple skirts. Increasing the NA does not always improve contrast.

12.6 100X Aplanat

In Figs. 12.9 and 12.10, a 100X aplanat is displayed from Smith.[24] The lens employs a finite tube length of 180 mm. Consequently, its focal length is near 1.8 mm. The front end comprises three aplanatic surfaces of the same configuration in Fig. 12.2. The magnification of the front end is approximately 3.5X. The two additional doublets provide another 30X. The doublets are comprised of CaF_2, F2, and BaF2. There is a small curvature toward the object field. Consequently, it is not a plan objective. Its lack of spherical aberration and coma define an aplanat.

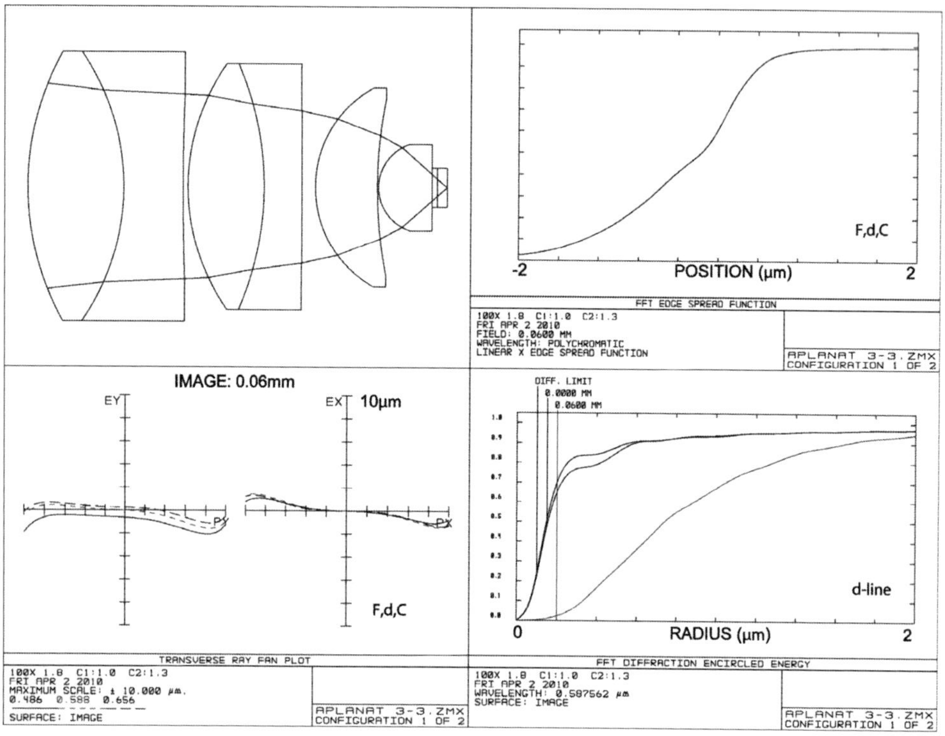

Figure 12.9 100X aplanat at 1.0 NA. NA_P = 1.3 is derived from a 3.5-mm entrance pupil and a 1.8-mm EFL at the d line. Adapted from Smith.[24]

In Fig. 12.9, the central NA is 1.0. The on-axis point displays a small fifth-order spherical aberration. This is caused by small differences in refractive index between the sphere BK7, the oil TYPE A, and the cover K7. These small changes in index at plano interfaces create spherical aberration. In Fig. 12.10, the paraxial NA is 1.3. The fifth-order spherical aberration is large.

12.7 10X Schwarzschild

Figure 12.11 displays a 10X Schwarzschild based on geometric relations in Reidl.[25] The focal length is 20 mm. The marginal NA is 0.25, and the central NA is 0.117. In the ray-intercept plot, correction at two wavelengths (588 and 656 nm) indicates an achromat; lack of linear slope indicates focus on the curved object surface. The encircled-energy plot indicates nearly diffraction-limited performance at 0.6 mm and slightly aberration-limited performance at 0.6 mm. The flat of the encircled-energy plot indicates an Airy radius of 1.2–1.4 µm. The paraxial NA 0.25 indicates an Airy radius of 1.4 µm at 588 nm. The edge spread displays a high-contrast transition over 1.0 µm.

The efficiency of annular collection is based on the marginal and central NAs:

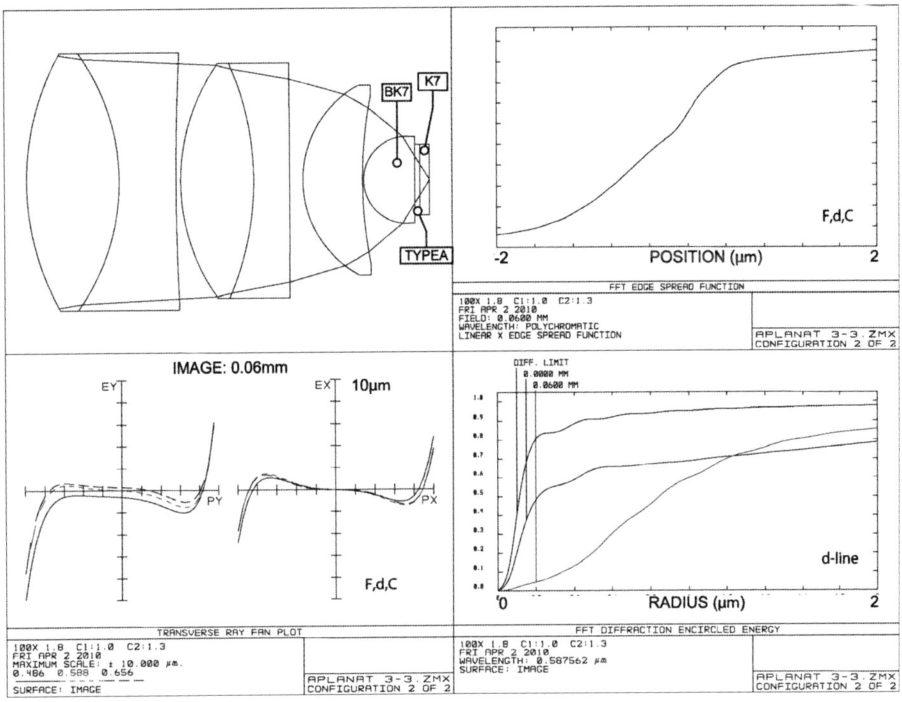

Figure 12.10 100X aplanat at 1.3 NA. $NA_P = 1.3$ is derived from a 4.5-mm entrance pupil and a 1.8-mm EFL at the d line. Adapted from Smith.[24]

$$\eta_{AC} = \sin^2\left[\frac{\arcsin\left(\mathrm{NA}_M/n\right)}{2}\right] - \sin^2\left[\frac{\arcsin\left(\mathrm{NA}_C/n\right)}{2}\right]. \tag{12.2}$$

The efficiency of annular collection in the current Schwarzschild is 1.2% in air and 0.7% in water.

12.8 20X Internal Parabola

Figure 12.12 displays a 20X internal parabola after Larson et al.[26] and Krogmeier et al.[27] The focal length is 10 mm. The marginal NA is 1.00. In the ray-intercept plot, a large amount of coma exists. The encircled-energy plot indicates diffraction-limited performance at 0.3 mm and aberration-limited performance at the 0.3-mm field. The geometric radius of the encircled-energy plot is 250 µm. At 20X magnification, this spot radius becomes a 5.0-mm-diameter spot, which provides a generous tolerance for placement of a ¾-in. photomultiplier tube. The SCE is 17% in water. An illuminator with a central NA of 0.25 defines an annular collection efficiency of 16.1% in water.

An issued patent for the internal parabola[26] provides a short tutorial on valid claims. The patent claims a chip comprising a mircofluidic channel, an illuminator, and a concave collector. Individually, any one of these items does

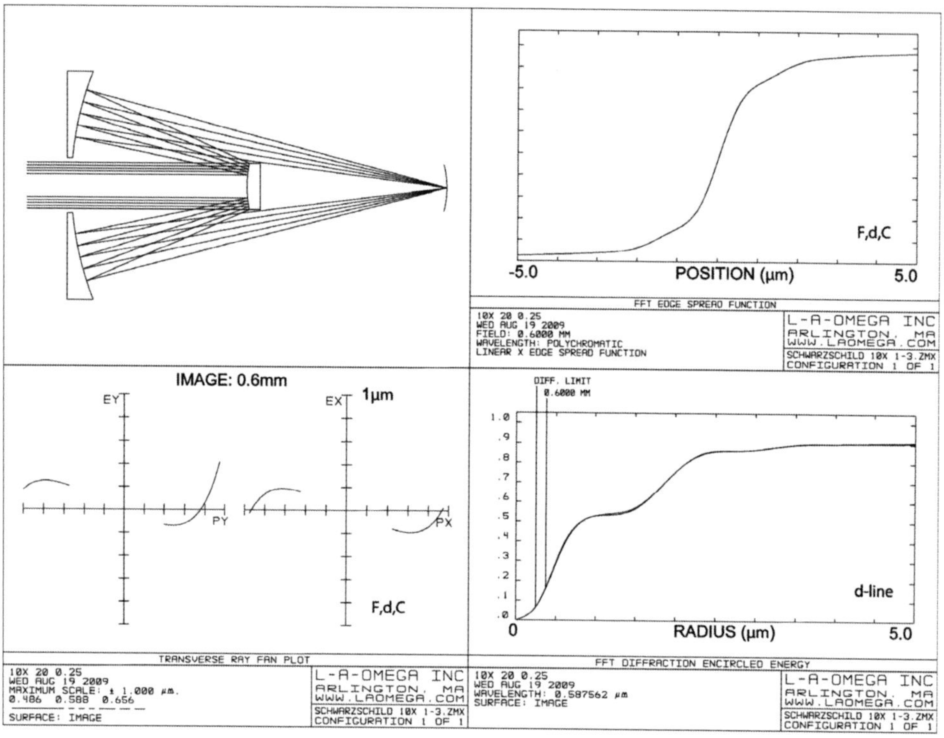

Figure 12.11 10X Schwarzschild at 0.25 NA. NA_M = 0.25 is derived from a 10.0-mm entrance pupil and a 20-mm EFL. NA_C = 0.117 is derived from a 10-mm entrance pupil and a 44.7-mm distance from convex mirror to field.

not comprise an invention; however, their combination does constitute an invention. A parabolic reflector by itself is certainly prior art from nearly a century ago. However, there are numerous patents on combinations of a parabolic reflector and a refractor. A signal lamp for railroads[28] employs a central refracting lens and an annular internal parabolic reflector as an improvement over metallic parabolic reflectors. The internal reflection of the annular collector is impervious to corrosion. Addition of several annular refractors greatly reduces the diameter of the parabolic reflector.[29] An LED mounted to the focal plane of a concave mirror comprises a novel combination.[30] An annular parabolic cap[31] collimates exiting light from the cylindrical surface of the LED lens. A cylindrical void defines a collection angle beyond a hemisphere within an annular reflector and a central refractor.[32] A cylindrical void contains a convex refractor, while the central refractor and the annular reflector share a planar exit face.[20] There are numerous combinations of parabolic reflectors and refractors.

The surface texture of the internal reflector is critical. At an index of 1.50, an internal reflector requires a surface texture 36 times smaller during refraction. An internal reflector at visible wavelengths has challenging tolerances for surface texture.

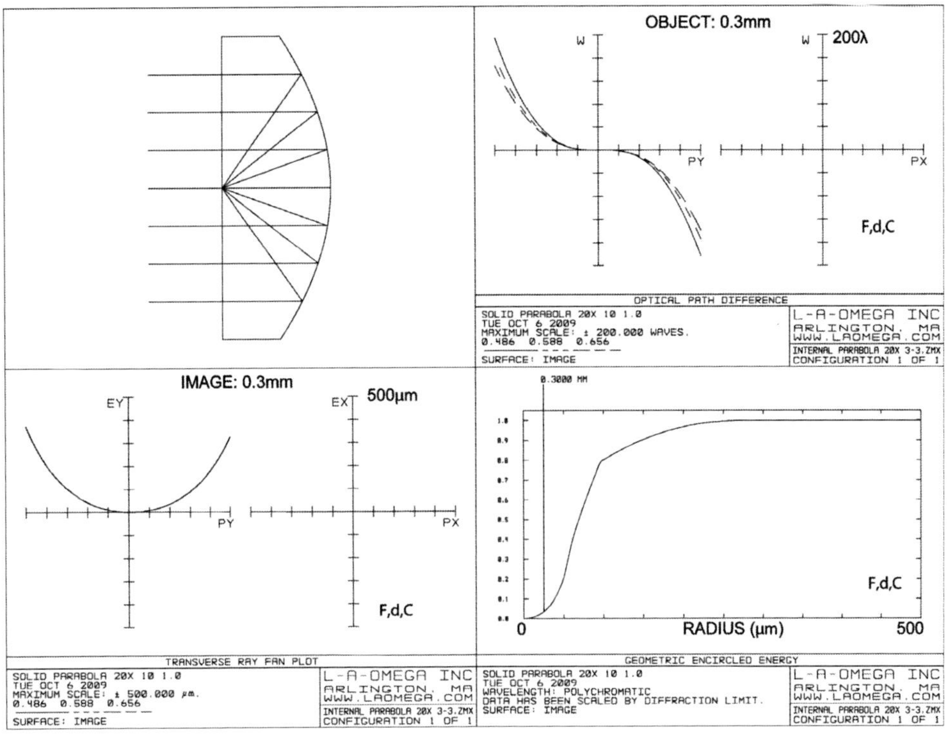

Figure 12.12 20X solid parabola at 1.00 NA_P. NA_M = 1.00 is derived from a 30-mm entrance pupil and a 10-mm EFL.

Chapter 13
Tube Elements

13.1 Doublet Tube Lens

A stock 200-mm doublet may serve effectively as a tube lens as shown in Fig. 13.1. The performance on axis is diffraction limited. At a 6.0-mm image field, there are several aberrations. The ray intercepts of the *x* plane indicate defocus in the form of Petzval curvature. The ray intercepts of the *y* plane indicate astigmatism and coma. The bulk of the aberration is field curvature. Consequently, the production doublet is effective within a small field with a CCD sensor.

A custom 200-mm doublet may provide a tube lens with flat field (Fig. 13.2). The performance on axis is diffraction limited. At a 6.0-mm image field, there is no field curvature. However, the ray intercepts of the *y* plane indicate a small coma. The encircled-energy plot indicates nearly diffraction-limited performance. A telecentric lens stop is essential to this design. The custom doublet is an effective design for a CCD sensor with a 12-mm diagonal.

13.2 Doublet-Pair Tube Lens

A typical infinity-corrected microscope employs a doublet pair as shown in Fig. 13.3. The first doublet is plano convex. It provides three features: optical power, correction of spherical aberration, and correction of axial color. The second doublet is a meniscus lens with little optical power. It provides correction of coma and lateral color. The "bending" of the second lens from flat to curved corrects the coma of the first lens.

The doublet pair based on the Abbe number provides excellent correction of geometric aberration (Fig. 13.4). The performance on axis is diffraction limited at the d line. There is only axial color. At a 6.0-mm image field, there is no field curvature. However, the ray intercepts of the *y* plane indicate a significant lateral color for the g line (426 nm). The encircled-energy plot indicates nearly diffraction-limited performance at the d line at a 6.0-mm field height. A telecentric lens stop is essential to this design. The doublet pair is an effective solution for a CCD sensor with a 12-mm diagonal.

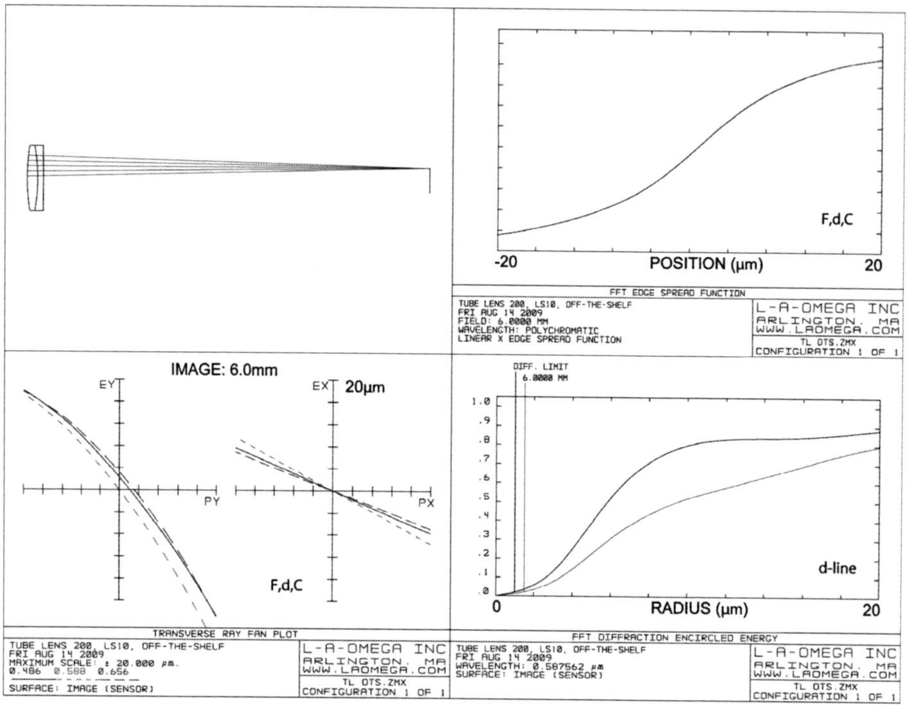

Figure 13.1 Stock doublet acts as a 200-mm tube lens. Lens stop is 10 mm in diameter. A telecentric lens stop is created by placement of a lens stop at the front focal point.

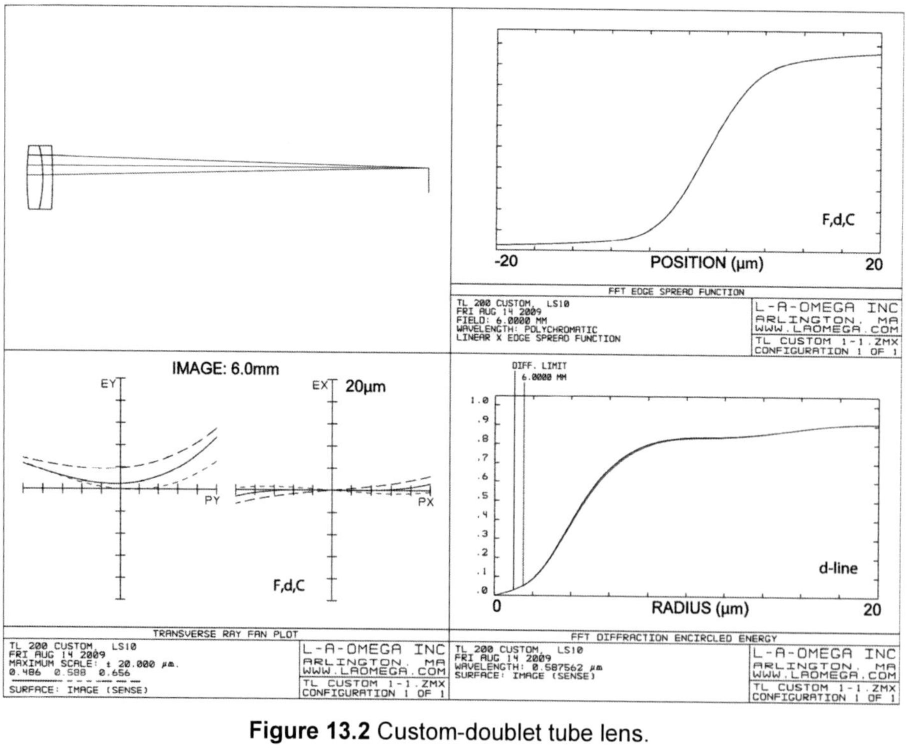

Figure 13.2 Custom-doublet tube lens.

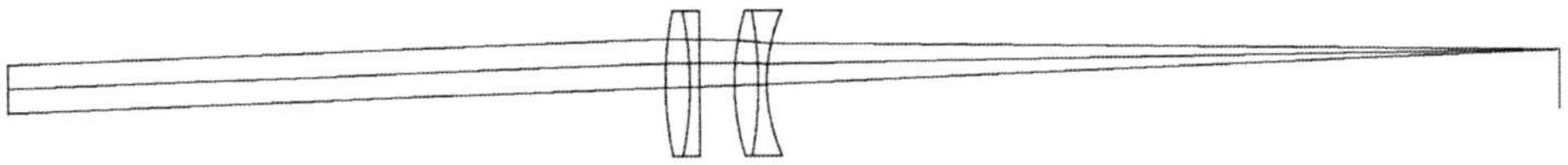

Figure 13.3 Doublet-pair tube lens layout. A custom doublet acts as a 200-mm tube lens. The lens stop is 10 mm in diameter. A telecentric lens stop is created by placement of a lens stop at the front focal point.

The doublet pair based on the Abbe number (shown in Fig. 13.4) may be considered an achromat for the d and g lines. Figure 13.5 displays the axial color of the doublet pair. The "U shape" indicates complete correction for two wavelengths. However, the plot does not display the "S shape" of an apochromat, where a vertical line indicates three corrected wavelengths. The lens is certainly an achromat; however, it is not apochromatic. The doublet pair might be called a semiapochromat because it is substantially corrected for three wavelengths.

Implementation of real glass from the 1990s yields improved performance of the doublet pair as shown in Fig. 13.6. The chromatic correction is greatly improved by application of known glass types to the prescription. The g line displays less lateral color. However, there is a cost to correct lateral color. The lanthanum flint LaFN7 and the short-flint special KzSFN4 are highly reactive with the atmosphere and cleaning agents. The schwer krone SK10, the barium schwer flint BaSF4, and the short-flint special KzSFN4 are obsolete as of 2010. Implementation of this design requires new glass types along with new values for radius and thickness. See Chapters 17 through 19 for additional information on glass types and correction of lateral color.

The doublet pair with real glass in Fig. 13.6 should be considered an achromat. Figure 13.7 displays the axial color of the doublet pair with real glass. The lens with real glass is barely an achromat. However, the doublet pair may be called a semiapochromat because it is substantially corrected for three wavelengths.

13.3 Filter Types

There are numerous types of filters: a long-pass filter transmits long wavelengths; a short-pass filter transmits short wavelengths; a bandpass filter transmits a narrow band; a dichroic mirror typically reflects shorter wavelengths; a notch filter reflects a narrow wavelength range.

Most filters comprise a dielectric stack on a 3-mm-thick substrate of fused silica. According to Eq. (5.35), the filter wavelength displays a dependency on the angle of incidence:

$$\lambda'_F = \frac{\lambda_F}{\cos\theta}, \tag{13.1}$$

where λ_F is the filter wavelength at normal incidence and λ'_F is the equivalent wavelength at an incident angle θ.

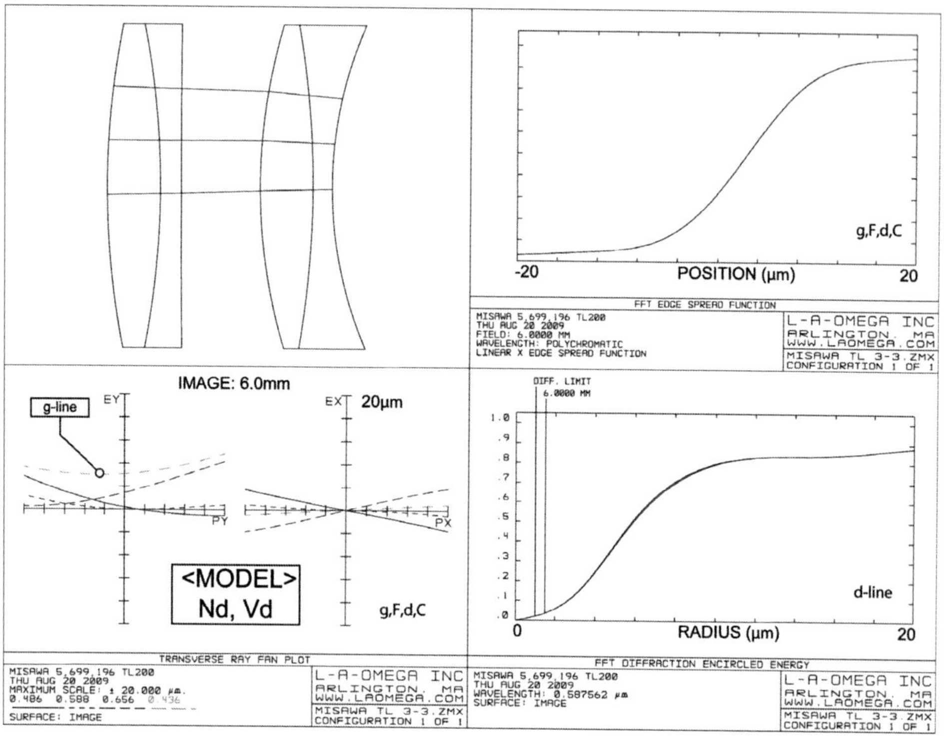

Figure 13.4 Doublet-pair tube lens based on the Abbe number. The g line indicates lateral color. Glass is based on numeric values of the refractive index and the Abbe number at the d line.

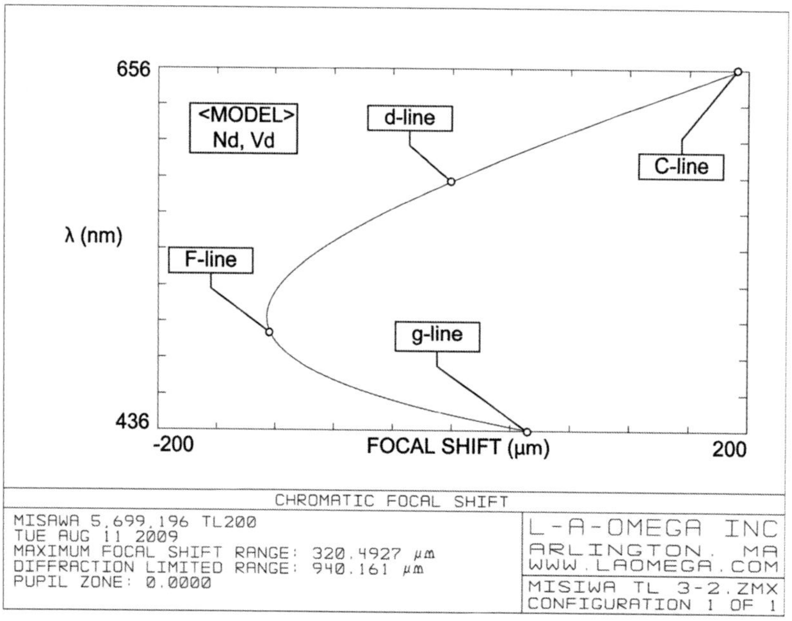

Figure 13.5 Axial color of a doublet-pair tube lens based on the Abbe number. A custom doublet acts as a 200-mm tube lens. The "U shape" indicates an achromat.

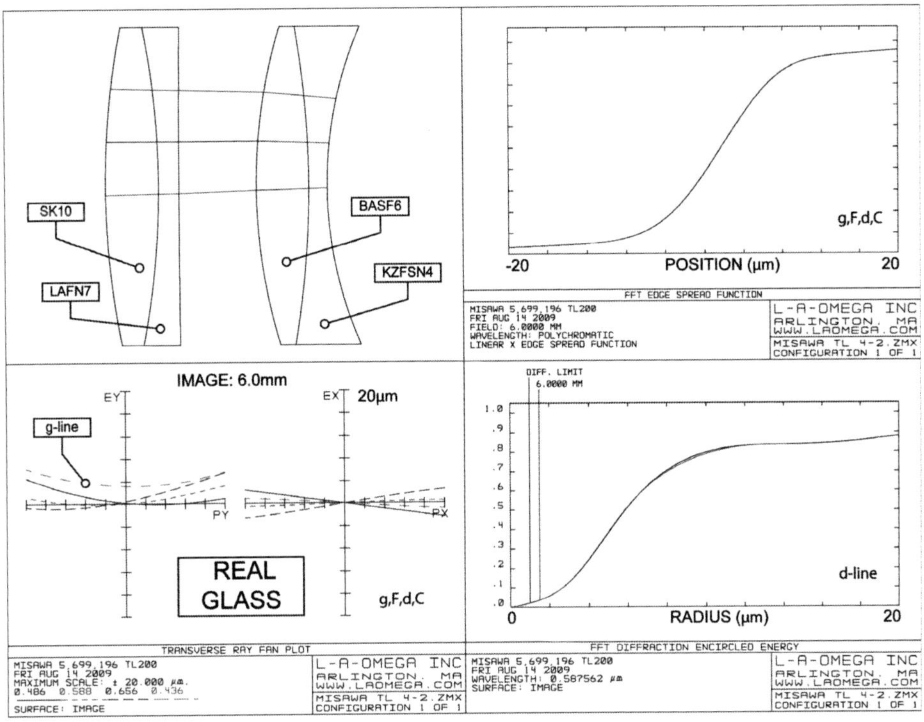

Figure 13.6 Doublet-pair tube lens based on real glass. Application of real glass yields better color correction. However, all but one of the glass types is obsolete as of 2010.

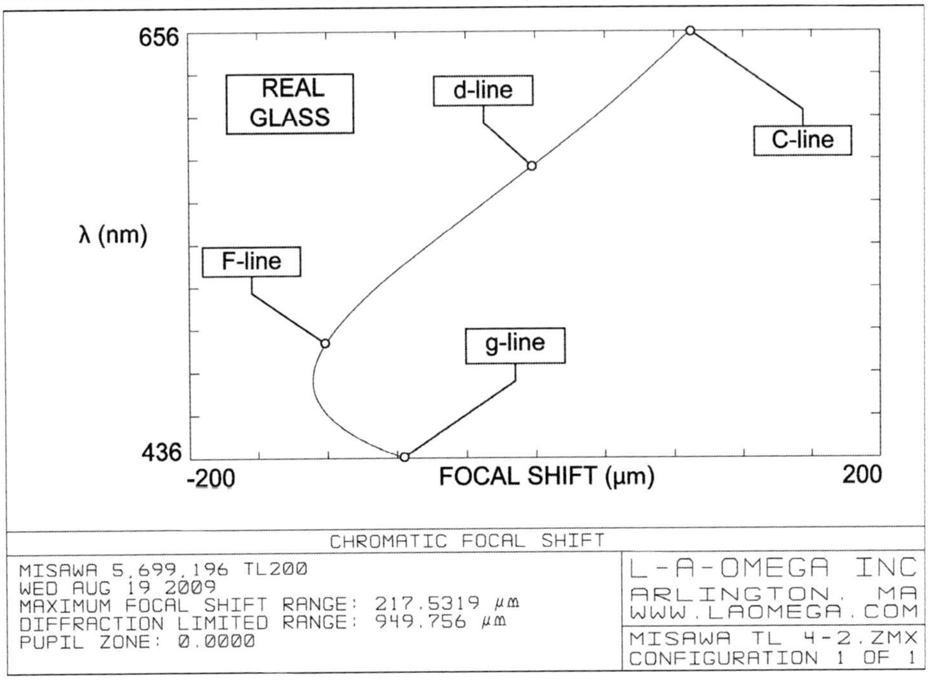

Figure 13.7 Axial color of a doublet-pair tube lens based on real glass. A custom doublet acts as a 200-mm tube lens. The "U shape" indicates an achromat.

Filter cubes employ thin 1.1-mm thick beamsplitters that reduce the displacement of the beam to < 1 mm. However, they are notorious for warping. The surface tension of the dielectric stack may deform a round filter into a spherical surface. However, surface tension frequently deforms a rectangular filter far beyond a sphere. The effects of deformation are small in transmission, but huge in reflection. A flatness specification is warranted for a filter, especially when the reflection of a filter is part of the imaging system.

A colored-glass filter employs absorption for transmission of long wavelengths. A colored-glass filter is independent from angle of incidence. However, colored-glass filters do emit fluorescence: absorption of a green laser beam by a colored-glass filter may create an orange emission from the filter.

A hard filter is comprised exclusively of metal oxides on a silica substrate. Numerous layers define the shape of the transmission spectrum. They are extremely durable during exposure to the atmosphere. Hard filters are a development from wavelength multiplexing in fiber-optic communications. The dielectric films are deposited under vacuum on a large substrate and then cut into smaller sizes.

A soft filter comprises two filters. The first filter is a short dielectric stack on the first substrate. The first filter provides multiple peaks in transmission. The second filter is a colored glass that absorbs the unwanted peaks in transmission. Absorption by the second filter frequently converts to fluorescence. The first and second filters are bonded to each other inside a ring, which provides a barrier against attack by the atmosphere. A soft filter may have a significant wedge that shifts the image location. There is a preferred direction of travel in a soft filter, because the dielectric filter should be the incident filter. Otherwise, the absorption filter encounters excess light, which promotes damage to the absorption filter.

Neutral density (ND) filters come in two types. An adsorptive ND filter absorbs light throughout the volume of the filter. It appears gray or black. A reflective ND filter employs a thin film of inconel for both reflection and absorption. Inconel is an alloy that is extremely resistant to oxidation and corrosion. Its primary composition is nickel (72%), chrome (15%), and iron (10%).

Filters are normally specified by optical density (OD), as follows:

$$\mathrm{OD} = -\log_{10} T, \tag{13.2}$$

where T is the transmittance. Thus, an optical density of 3 indicates a transmission of 0.001. A small optical density of 0.05 adds linearly to another small optical density, such as 0.10. However, large optical densities do not necessarily add together due to large reflections.

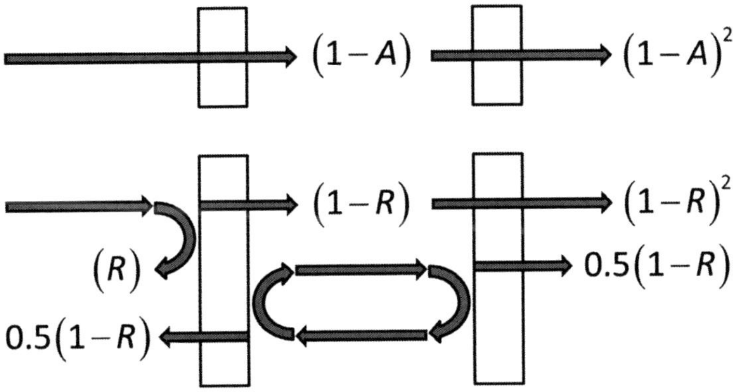

Figure 13.8 Filter doublet. An absorbance indicates an absorption filter without reflection (*A*). *R* indicates a reflection filter without absorption. Light within a high-reflectance cavity has equal probability of exiting through either filter.

Figure 13.8 displays a filter doublet wherein light may circulate in the cavity between the filters. An absorption filter is described by absorbance A, while a reflection filter is described by reflectance R. A sequence of two absorption filters yields the square of the transmittance values as the total transmittance. Consequently, their optical densities add to the total optical density. A sequence of two reflection filters yields far more than the square of the transmittance values. Addition of the second filter cuts the transmittance in half. Consequently, the optical densities of two reflective filters do not add to the total optical density. They might even disrupt each other if the separation is less than the coherence length of the transmitted spectrum.

13.4 Filter within a Finite Conjugate Distance

A perfect tube conjugate is defined by a 10-mm stop and a 200-mm focal distance. This defines a tube NA at 0.025, which is rather small. It can tolerate a large thickness of cover strata without creation of spherical aberration. However, it cannot tolerate much thickness for a tilted plate.

Figure 13.9 displays a 50-mm beamsplitting cube in perfect tube conjugate. The ray intercept for the x plane indicates axial color. The ray intercept for the y plane indicates both axial and lateral color. A 50-mm beamsplitting cube is a reasonable limit for a 0.025 NA conjugate. This represents a wavefront error of 0.26λ.

Figure 13.10 displays a 1.4-mm beamsplitting plate in perfect tube conjugate. The ray intercept for the x plane indicates proper focus. The back focal length is driven to proper focus with the x plane. The ray intercept for the x plane indicates astigmatism and lateral color. A 1.4-mm beamsplitting plate is acceptable within a 0.025 NA conjugate. This represents a wavefront error of 0.34λ.

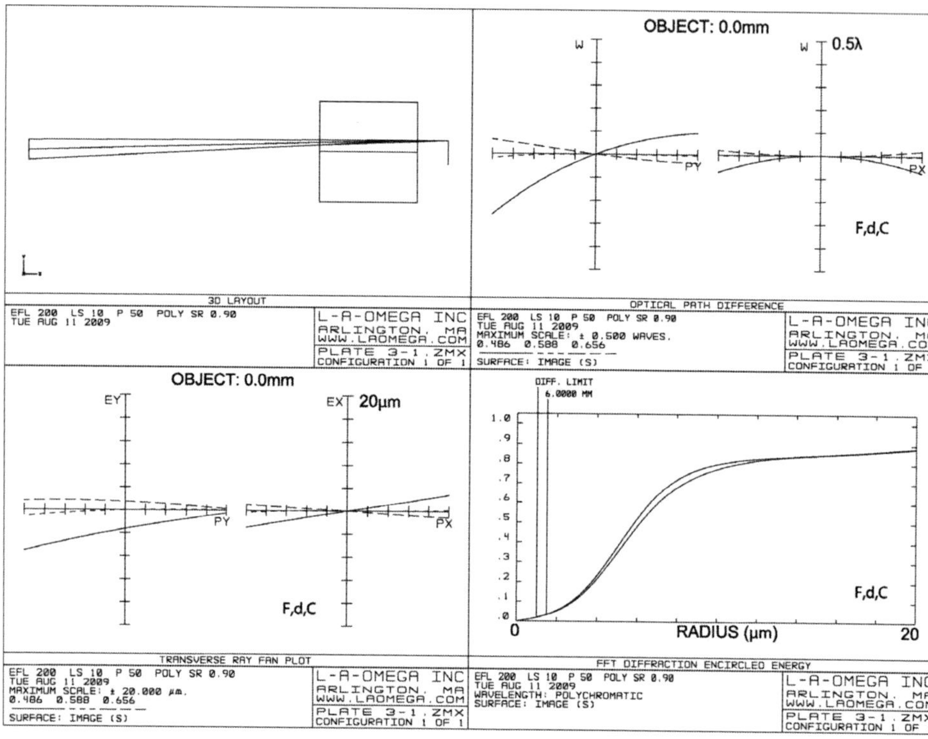

Figure 13.9 Beamsplitting cube within a finite conjugate distance. The cube consists of 50 mm of silica. The polychromatic Strehl ratio is 90%. A perfect lens is defined by a 10-mm lens stop and a 200-mm EFL, and NA_P = 0.025. Exception from other figures: encircled energy applies to a polychromatic combination of F, d, and C lines.

13.5 Warped Filter within an Infinity Correction

A warped plate may also create astigmatism in a collimated beam of the infinity-corrected tube. Two common filter thicknesses are evaluated: 30 and 1.1 mm.

A tilted plate creates a lateral displacement of the beam:

$$\Delta y_P = d_P \sin\theta \left(1 - \sqrt{\frac{1-\sin^2\theta}{n^2-\sin^2\theta}} \right). \tag{13.3}$$

In fused silica at 45 deg, the lateral displacement is 0.32 times the plate thickness. A 3.0-mm thickness creates a 1.0-mm beam shift, while a 1.1-mm thickness creates a 0.4-mm beam shift.

In Fig. 13.11, a tilted filter at 3.0 mm in thickness is located within a tube of an infinitely distant conjugate. A warp radius of 580 mm creates an astigmatism at a Strehl ratio of 0.9. This represents a peak wavefront error of 0.40λ within a 10-mm lens stop.

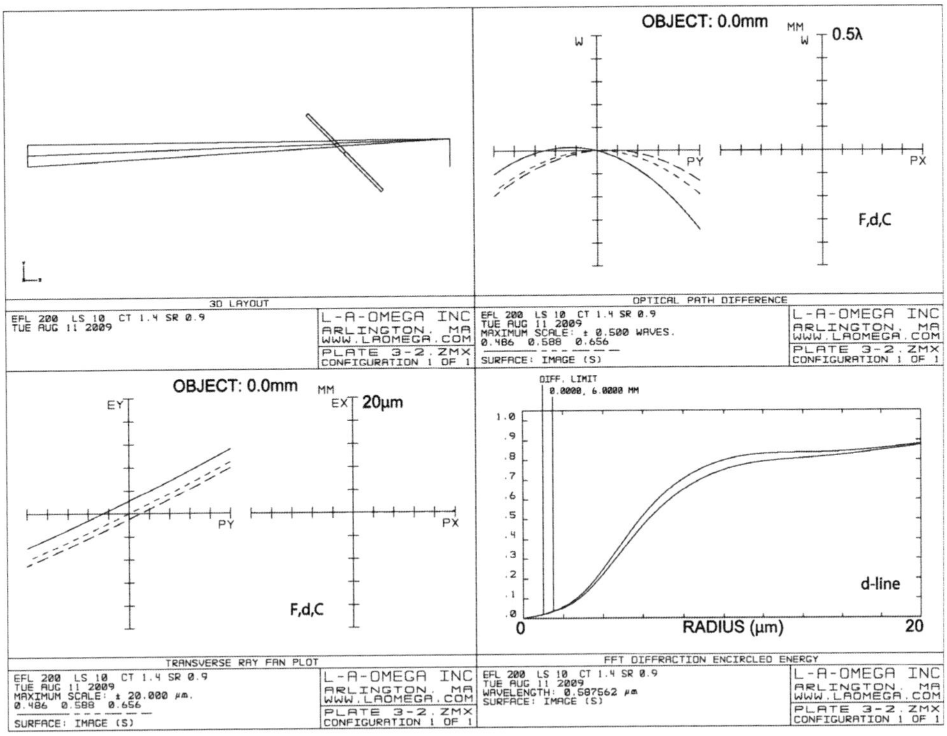

Figure 13.10 Plate tilted within a finite conjugate distance. A parabolic reflector defines the tube lens as a 10-mm lens stop and a 200-mm EFL. The plate at a 45-deg tilt consists of 1.4 mm of silica. The Strehl ratio of the d line is 90%. A perfect lens is defined by a 10-mm lens stop and a 200-mm EFL, and the $NA_P = 0.025$.

In Fig. 13.12, a tilted filter at 1.1 mm in thickness is located within a tube of an infinitely distant conjugate. A warp radius of 350 mm creates an astigmatism at a Strehl ratio of 0.9. This represents a peak wavefront error of 0.35λ within a 10-mm lens stop.

A 1.1-mm-thick beamsplitter is commonly used within a filter block of a microscope. At 1.1 mm in thickness, a 25 × 36-mm filter plate can warp significantly. The corners are susceptible to greater bending than the center. This creates additional astigmatism and defocus at the corners. A tolerance on radius of curvature or transmitted wavefront error is warranted.

The above examples employ a parabolic reflector as the tube lens. Each system is driven to the focus of the *xz* plane. The parabolic reflector is perfect for an on-axis object, while it is not suitable for an off-axis object. Consequently, an on-axis object is employed in the examples for an infinity-corrected tube. Ideally, a tolerance for a filter curvature should be calculated in combination with an actual tube-lens design.

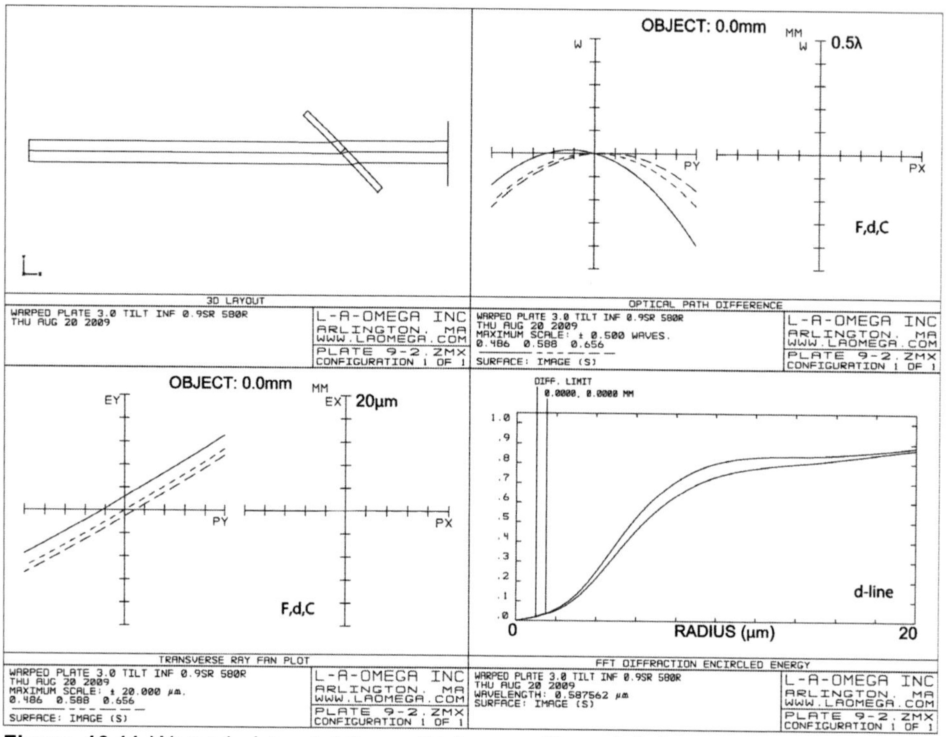

Figure 13.11 Warped plate at 3.0-mm thickness with an infinity correction. The warped plate at 45-deg tilt consists of 3.0 mm of silica. A 580-mm radius of curvature is derived from a Strehl ratio of the d line at 90%.

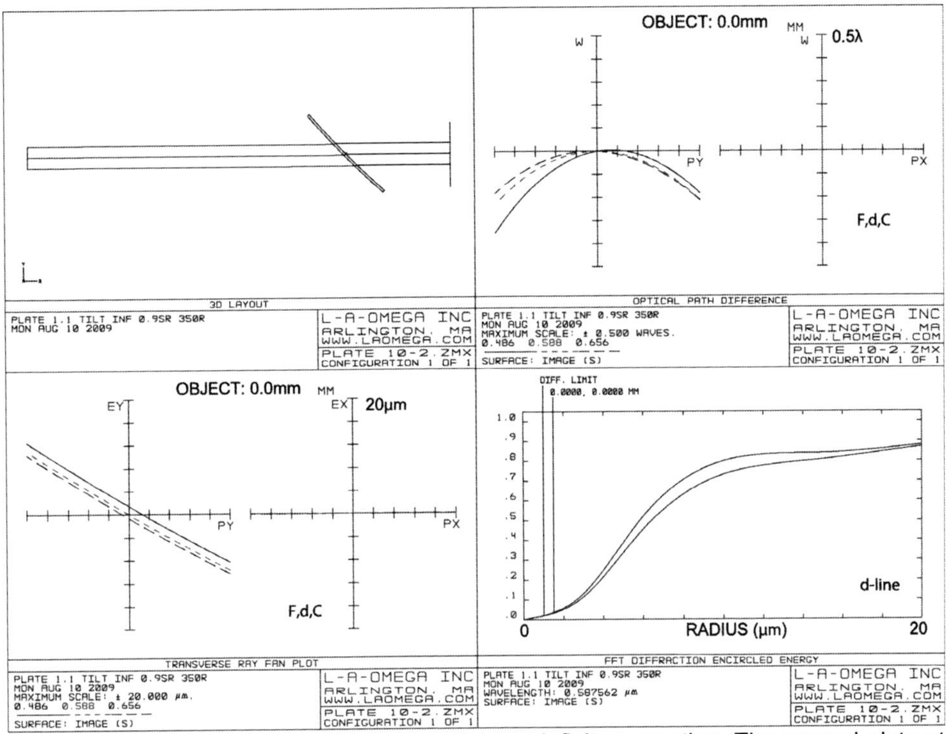

Figure 13.12 Plate at 1.1-mm thickness within an infinity correction. The warped plate at 45-deg tilt consists of 1.4 mm of silica. A 350-mm radius of curvature is derived from a Strehl ratio of the d line at 90%. A tube lens is a parabolic reflector with a 10-mm lens stop, 200-mm EFL, and $NA_P = 0.025$.

Chapter 14
Ocular Lenses

14.1 Eyepiece

An eyepiece is a magnifier with its entrance pupil defined by the lens stop of an objective lens. It is also known as an ocular lens. A typical ocular has a focal length of 25 mm, which is one-tenth of the near point of human vision at 250 mm. An image conjugate of the tube lens serves as the object of an ocular. The object is located at the front focal point of the eyepiece. The angular size of the image is 10 times greater than the angular size of the object at the standard near point. Consequently, the angular magnification of a typical eyepiece is 10X. The angular magnification of an arbitrary eyepiece is defined by Eqs. (2.7) and (3.27).

14.2 Pupils

The entrance pupil of an eyepiece is defined by the lens stop of the objective lens. In a system with finite tube length, the object NA is defined by the diameter of the objective lens stop and the length of the tube. In a system with infinite tube length, a tube lens creates a distant image of the objective lens stop, and the object NA is defined by the diameter of the lens stop and the focal length of the tube lens.

The exit pupil of an eyepiece is an image of the entrance pupil, which is the lens stop. During proper illumination, the exit pupil appears as a bright white disk floating above the eyepiece. The exit pupil should fit within the pupil of a human eye. The distance from the last surface of the ocular to the exit pupil is the eye relief.

The eye relief is an important functional parameter. At short distances of 10 to 15 mm, the eye relief prevents contact with eyelashes. At longer distances of 25 to 30 mm, the eye relief provides space for eyeglasses. A flexible eye cup on the ocular acts as a shield from room light for an operator without eyeglasses. The eye cup is folded backward along the ocular for operators with eyeglasses.

The diameter of the exit pupil defines an artificial pupil for the operator's eye. At small pupil values of 1.25 mm, the human eye is diffraction limited: this is true even for people with minor astigmatism. This artificial pupil provides a significant advantage over the natural pupil of 2.5 mm at which the eye is aberration limited. Proper room light is required for dilation to a natural size of 2.5 mm.

The field stop of the ocular is placed at an image field of the tube. The field stop of the ocular defines the vision field at the specimen. An image of the field stop may be projected onto the vision field. The size of the field stop is dependent on the type of ocular lens. The diameter of the field stop can vary from 15 mm in a Kellner ocular to 25 mm in an Erfle ocular.

In a microscope of finite tube length in 1980, an image is located at 160 mm from the objective flange—this defines a tube length of 160 mm with an image located at its end. The field stop of the ocular is located at the end of the tube. A field stop of 22 mm defines a chief ray angle of 3.9 deg in an eyepiece of 25-mm EFL. Beyond this angle, the image quality becomes more difficult to maintain. Consequently, the lens stop of a microscope of finite tube length is typically 22 mm or less.

In a microscope with infinity correction, a 10-mm lens stop and a 200-mm tube lens create a tube NA of 0.025. The field stop of the ocular is located at the back focal point of the tube lens.

14.3 Kellner Ocular

In Fig. 14.1, a Kellner eyepiece employs a doublet and a plano-convex singlet. The doublet acts as a magnifier on the image, while the singlet acts as a relay lens for the entrance pupil (which is the lens stop). The singlet is called a field lens because it is located near the image field, where it has little effect on the magnification of the field. Defects of the singlet can represent a significant portion of a ray bundle. Consequently, the singlet is placed near the image field but not at the image field.

In Fig. 14.2, a Kellner ocular lens is formed by two stock items, a plano-convex lens of 25-mm EFL and doublet of 30-mm EFL. The maximum field is limited to 7.5 mm in the field stop and 17 deg in the image. The collinear ray-intercept plots of the x plane indicate a nearly perfect spot on axis with a small Petzval curvature. The ray-intercept plots of the y plane indicate axial color, lateral color, and astigmatism. The field-stop diameter is limited to 15 mm by the outer diameter of the doublet.

Figure 14.1 Kellner ocular at a finite distance to the lens stop. The lens stop (LS) is 8 mm in diameter. The tube length is 160 mm. The object NA is 0.025. The field stop (FS) is 15 mm in diameter. The exit pupil (EP) is 1.6 mm in diameter at an eye relief of 15.5 mm. the field of the image is 17 deg. The field lens (FL) is a standard plano-convex lens at 50 mm in focal length. The magnifier (M) is a stock doublet at 30 mm in focal length. The EFL is 25.0 mm at the d line. The angular magnification is 10X the near-point view.

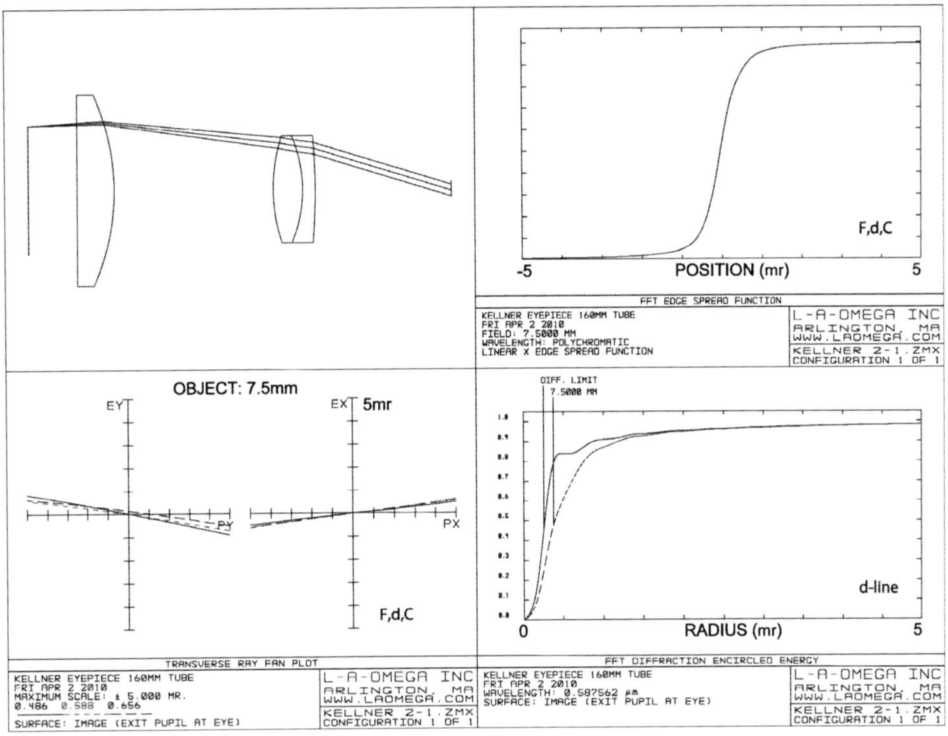

Figure 14.2 Kellner lens report. Wavelengths are at the F, d, and C lines. Object height is 7.5 mm in the lens report. Maximum field diameter is 15.0 mm.

A Kellner lens is an excellent low-cost option with perfect image quality on axis. The angular field at 17 deg and the field stop at 15 mm are both small by today's standards. Replacement of the doublet with a plano-convex singlet creates a Ramsden eyepiece, which is a predecessor to the Kellner.

14.4 Plössl Ocular

A Plössl eyepiece is formed by two modifications to the Kellner. The first modification involves splitting the doublets of a Kellner eyepiece into two doublets, and the second modification involves removing the field lens. Splitting the magnifier of the Kellner into two doublets greatly increases the diameter of the magnifier while maintaining its focal length. The increased diameter of the magnifier permits elimination of the field lens. A Plössl benefits from a telecentric object space because less bending of rays is required. A Plössl eyepiece is common in telescopes.

In Fig. 14.3, a Plössl ocular is formed by a pair of standard-stock doublets at 50 mm in focal length. The collinear ray-intercept plots of the *x* plane indicate near perfection on axis. The sloped ray-intercept plots of the *x* plane indicate field curvature. The ray-intercept plots of the *y* plane indicate astigmatism and lateral color. The field-stop diameter is limited to 25 mm by the outer diameter of the doublet.

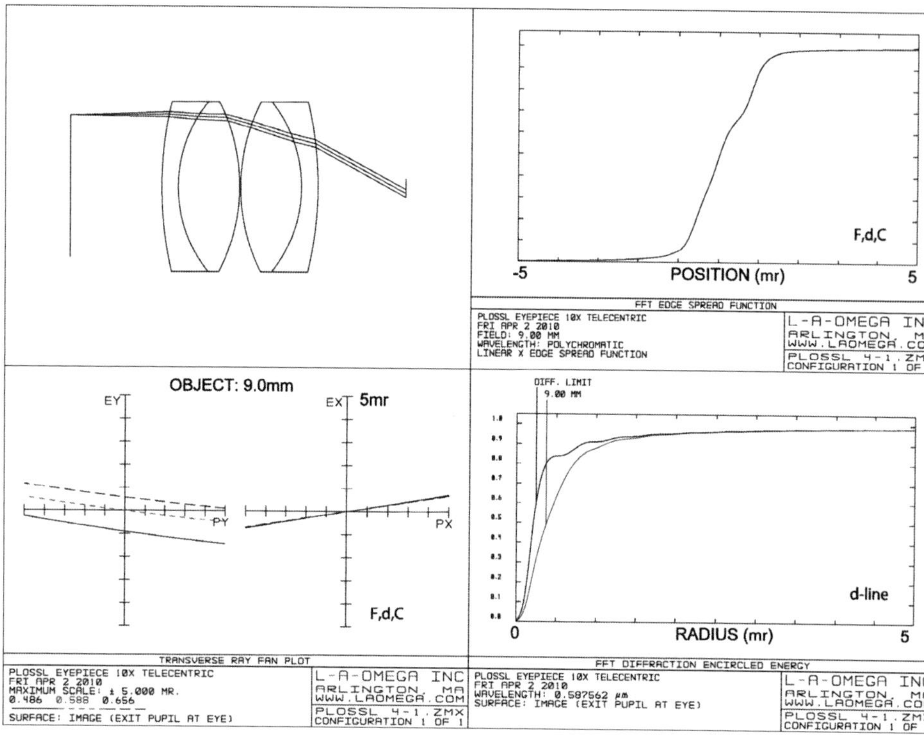

Figure 14.3 Plössl lens report. The lens stop is 50 mm in diameter at 1000 mm from the field stop. The object NA is 0.025. The exit pupil diameter is 2.0 mm at an eye relief of 19.0 mm. Doublets are identical stock products at 50 mm in focal length. The EFL of the doublet pair is 26.6 mm at the d line. The angular magnification is 9.4 times the near-point view. Object height is 9.0 mm in the lens report. Maximum field diameter is 25.0 mm.

A Plössl lens is an excellent low-cost option for a wide-field eyepiece. The angular field at 28 deg and the field stop at 25 mm are both large by today's standards. Glass optimization can reduce the lateral color.

14.5 Erfle Ocular

A historical development of the Erfle eyepiece is described by two prescriptions. The original patent by Erfle in 1923 is displayed in Fig. 14.4. It was based on the limited selection of available glass at the time. An Erfle eyepiece of 1997 is displayed in Fig. 14.5 with higher-index glass and low-dispersion lanthanum glass. The glass types of these Erfle lenses are displayed in the glass map of Fig. 14.6.

The Erfle design of 1923 employs glass from the lower region of the glass map: borosilicate krone (BK) and flint (F). The refractive index does not exceed 1.62. Consequently, the element curvatures are steeply curved, which enhances lateral color and astigmatism. The dispersion of these glass types does not exceed 36. Consequently, there is limited color correction at large field angles. The field-stop diameter is limited to 25 mm by the outer diameter of the elements.

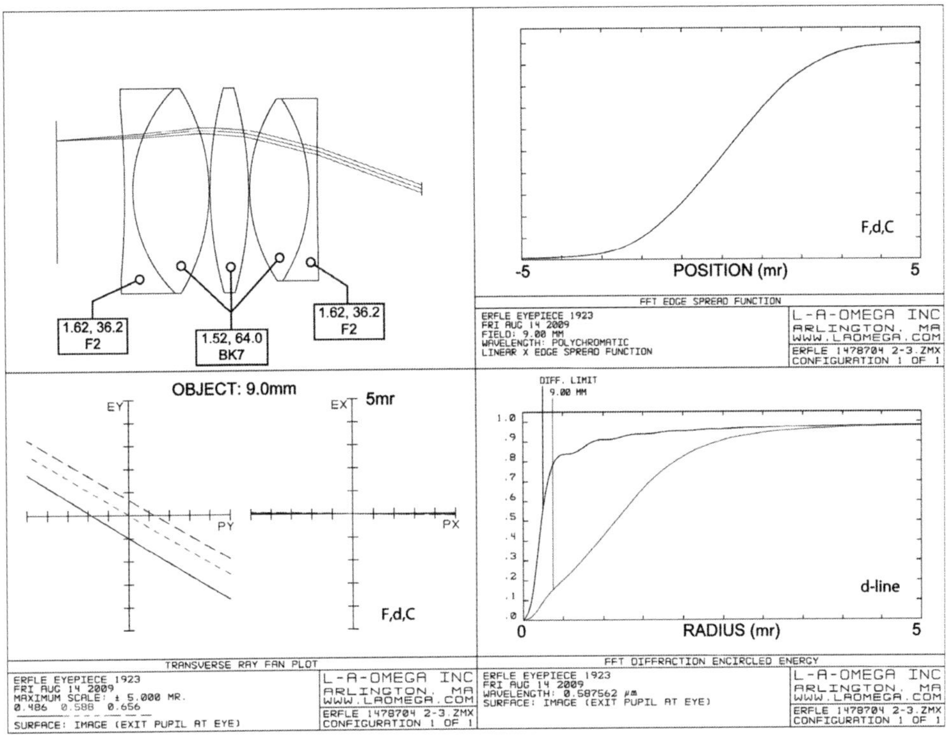

Figure 14.4 Erfle lens of 1923. The lens stop is 10 mm in diameter at 200 mm from the field stop. The object NA is 0.025. The exit pupil diameter is 2.3 mm at an eye relief of 18.7 mm. The elements are custom to the design. The EFL of the doublet pair is 25.2 mm at the d line. The angular magnification is 9.9X the near-point view. Object height is 9.0 mm in the lens report. Maximum field diameter is 25.0 mm.

The Erfle lens of 1923 is aberration limited. The collinear ray-intercept plots of the x plane indicate near perfection on axis. The flat ray-intercept plots of the x plane indicate no field curvature. The ray-intercept plots of the y plane indicate astigmatism. The encircled-energy plot indicates a spot radius of 3 mrad. The line-spread plot indicates a line width of 4 mrad.

The Erfle lens of 1997 employs glass from the upper region of the glass map, such as schwer flint (SF), schwer krone (SK), and lanthanum krone (LaK). The refractive indices greatly exceed 1.62. Consequently, the element curvatures are less steep, thus reducing astigmatism. The dispersion exceeds 36. Consequently, there is improved color correction at the field margin. A negative relative partial dispersion ΔP_{gF} of the lanthanum krone LaK8 provides correction of color in the blue. Additional information on glass types is found in chapters 17 through 19.

The Erfle lens of 1997 is nearly diffraction limited. The collinear ray-intercept plots of the x plane indicate near perfection on axis. The sloped ray-intercept plots of the x plane indicate a small field curvature. The ray intercept plots of the y plane indicate small lateral color. The difference in slopes between the x and y planes indicates a small astigmatism. The astigmatism and lateral

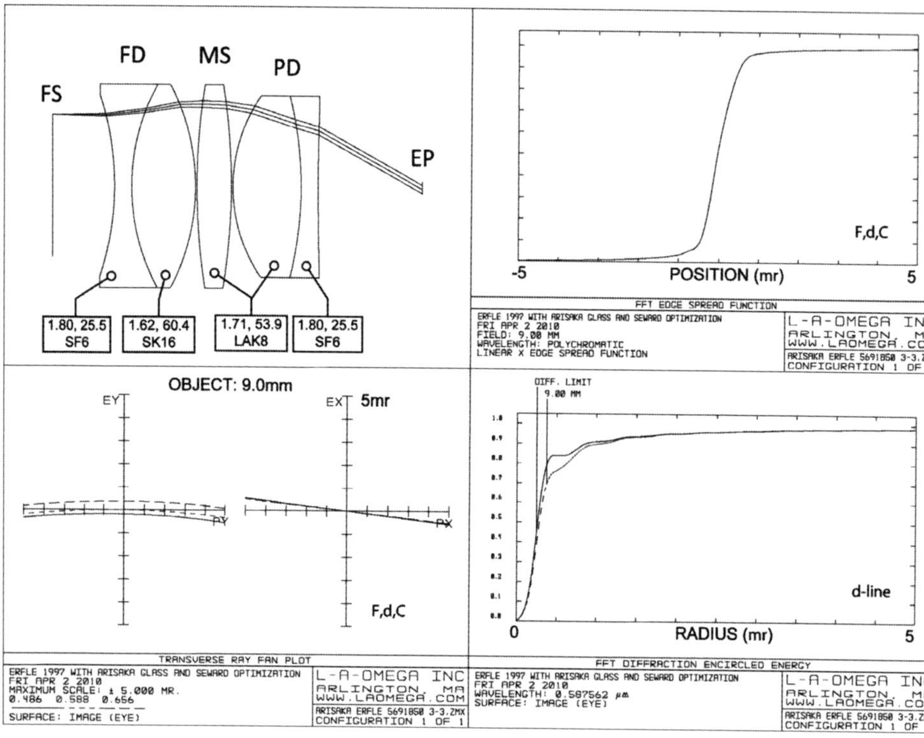

Figure 14.5 Erfle lens of 1997. The lens stop is 35 mm in diameter at 200 mm from the field stop. The object NA is 0.25 telecentric. The exit pupil (EP) is 2.1 mm in diameter at an eye relief of 18.8 mm. The field doublet (FD) and the pupil doublet (PD) are custom doublets. The middle singlet (MS) is a custom biconvex singlet. The EFL of the lens system is 29.0 mm at the d line. The angular magnification is 10.0 times the near-point view. Object height is 9.0 mm in the lens report. Maximum field diameter is 25.0 mm.

color are balanced in magnitude. The lens is substantially optimized. The flat of the encircled-energy plot indicates a spot radius of 1.0 mrad. The line-spread plot indicates a line width of 0.4 mrad.

The pupil doublet and the middle singlet contribute most of the optical power. The field doublet has little effect on the angle of the chief ray. At 9.0 mm in the object, the first doublet bends the chief ray by < 1 deg. The pupil doublet and the middle singlet contribute negative astigmatism due to their positive powers. The pupil doublet is substantially corrected for lateral color, while the middle singlet adds negative lateral color.

The field doublet effectively counters the aberrations of the pupil doublet and middle singlet. The biconcave flint of the field doublet creates a large positive lateral color. The steep convex surface of the field doublet creates a large angle of incidence for a chief ray. Consequently, the steep convex surface of the field doublet creates a positive astigmatism.

The distortion of the Erfle 1997 eyepiece is displayed in Fig. 14.7. A square box in the object field appears as a pincushion in the image field. This represents positive distortion that is caused by a negative growth in focal length with field height. The shorter focal length of the margin creates a greater magnification at the margin. A linear feature of the object is curved in the image field. An addition of a negative field singlet may reduce positive distortion.

Most of today's eyepieces are derivatives of the Erfle eyepiece. A Skidmore eyepiece splits the middle single into two singlets. An aspheric field lens may completely eliminate field curvature or distortion.

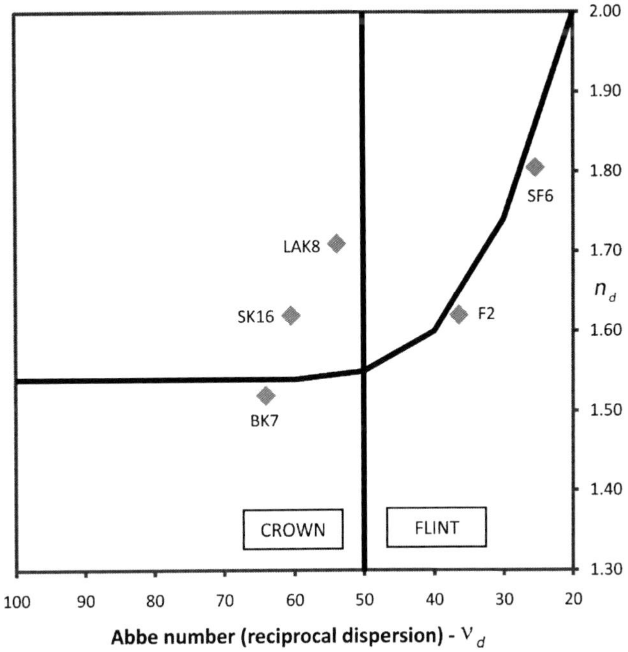

Figure 14.6 Glass map for Erfle oculars.

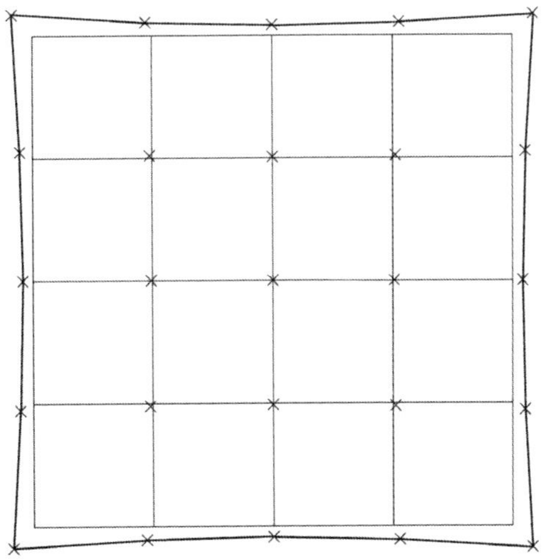

Figure 14.7 Grid distortion for the Erfle lens of 1997. The pincushion distortion is a positive distortion of 6.3% between the object and image. Wavelength is at the d line. Object height is 9.0 mm.

Chapter 15
Sensors

15.1 CCD Sensors

A CCD sensor is a charge-coupled device. Potential wells for electrons are created by extrinsic doping of a semiconductor. The metal-oxide gate of a field-effect transistor determines the depth of a well. An external voltage may raise or lower a gate. A specific gate sequence transfers electrons between wells. The charge-coupling process converts photoelectrons into a video signal. Holst and Lomheim,[33] provide a broad review of CCD and complementary metal-oxide semiconductor (CMOS) sensors. Janesick[34] provides a more rigorous review of CCD technology.

Figure 15.1 displays the structure of a potential well of a CCD sensor. An n-type material donates negatively charged electrons. A p-type material donates positively charged holes. The bulk material is p-type silicon. The Fermi level defines an equal probability for a hole and an electron. n-implantation raises the Fermi level by donation of electrons. Consequently, the conduction-band edge is lower in the n-doped region. This defines the base of the well. The full-well

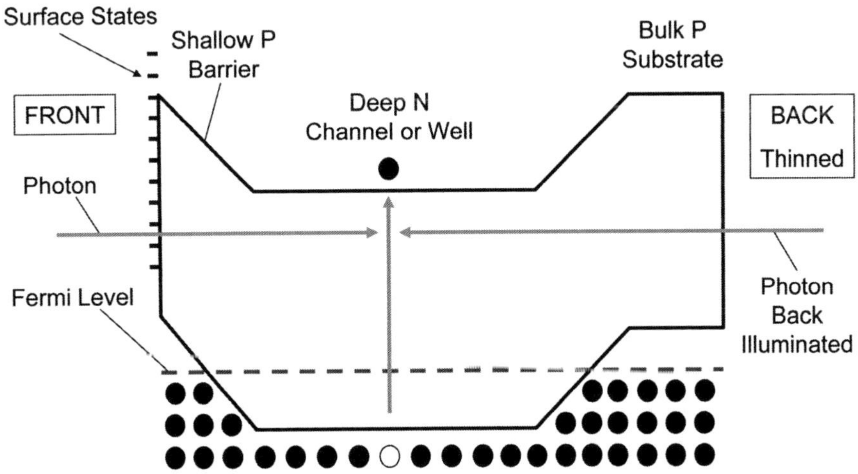

Figure 15.1 Potential well of a CCD. Photon absorption promotes electrons (filled circles) from valence band to conduction band. N-doping of bulk p-substrate creates a potential well for photoelectrons. Shallow p-doping isolates the potential well from surface states. A buried channel is defined by deep n-implantation and shallow p-implantation.

capacity is proportional to the depth of n-implantation. A deep well is defined by deep implantation of n-type atoms. The extinction depth of silicon is approximately 100 nm at 500 nm in wavelength. The well should exceed this depth for maximum quantum efficiency of conversion of photons into electrons.

A buried channel is formed in two steps: deep implantation of n-type atoms and shallow implantation of p-type atoms. Deep n-implantation requires a small-mass ion with a high energy. A small-mass ion loses a small fraction of its momentum during a collision with the base network of large-mass elements. Nitrogen and phosphorous are low-mass n-type ions. A high-energy ion is created by a particle accelerator with a high voltage. Shallow p-implantation requires a large-mass ion with a small energy. A large-mass ion element quickly transfers momentum to other large-mass elements of the base network. Gallium and indium are p-type ions with a large mass.

In summary, a buried channel requires two specific fabrication techniques: a high-energy low-mass n-implantation and a low-energy high-mass p-implantation. These features are becoming less common in silicon foundries, which produce CMOS devices such as flash memory and cell-phone cameras.

The metal gates occupy a significant portion of the pixel area. Consequently, the quantum efficiency of a front-illuminated CCD may be only 10%. A back-thinned CCD permits back illumination without obscuration by electrical contacts. The quantum efficiency of a back-illuminated CCD may exceed 90%; however, removal of the back surface is an expensive process. The back surface must be thinned to much less than the extinction depth.

Figure 15.2 displays a CCD circuit. Each field-effect transistor (FET) defines a well. The voltage of the metal-oxide gate may raise the bottom of the well. The raised gate drives photoelectrons into an adjacent well. A column is defined by a vertical array of pixels, where each pixel holds a packet of electrons. A clock sequence drives a packet along a column into a shift register. Another clock sequence transfers a row of packets to a sense node. The sense node converts the packets of charge into an analog or digital video signal.

There are two important types of noise from a CCD: dark current and read noise.[34] The dark current is created by thermal promotion of carriers into the conduction band. Surface states can randomly eject electrons as spurious charge or popcorn noise. A buried channel is essential for reduction of spurious charge from surface states. The read noise is created during amplification at the sense node. The thermal noise of the load resistor is typically the main source of read noise. Numerous types of noise exist within a CCD sensor.[34]

A Kodak KAI-1020 provides an example of noise in a CCD sensor at 40° C. Its pixel dimension is 7.4 μm. The full-well capacity of the 7.4-μm pixel in the KAI-1020 is 40,000 electrons. The dark current is typically 0.2 nA/cm^2, which translates to 600 electrons per pixel per second. The typical read noise is 10 electrons per pixel per second. At a frame rate of 30 Hz, the dark current contributes 20 electrons toward a total dark noise at 30 electrons. These specifications are readily available from the Kodak device specifications for the KAI-1020.[35]

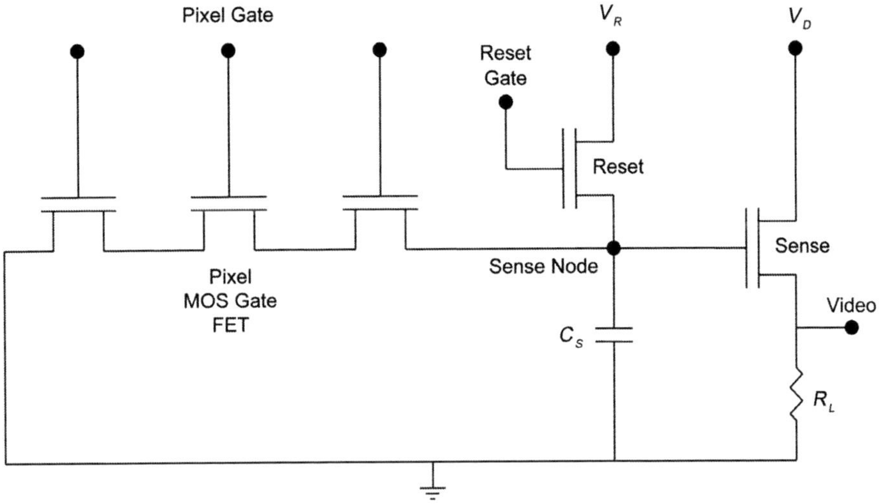

Figure 15.2 Circuitry of a CCD. A pixel is defined by a metal-oxide semiconductor (MOS) gate on a field-effect transistor (FET). A gate sequence delivers an electron packet to a capacitor at the sense node. The sense FET converts the packet to a video signal. Drive voltage (V_D) sends current through the sense FET and load resistor (R_L). A reset gate restores the sense node to the reference voltage (V_R).

The integration of dark current can be reduced by cooling or electron multiplication. Thermoelectric (TE) cooling may greatly reduce both read noise and dark current; however, TE cooling requires a dry enclosure for the sensor. Electron multiplication by a gain register before the sense node may quickly amplify the electron count without integration of dark current. An electron multiplication CCD (EMCCD) eliminates dark current at a cost of spurious charge. An EMCCD is an attractive alternative to TE cooling, which is more common in optical instruments.

A third type of noise is the shot noise of the photoelectrons. The full-well capacity of an 8-μm pixel is typically 40,000 electrons as the saturation signal.[35] This requires a 16-bit digital number (65536) for counting every photoelectron. However, the shot noise is the square root of the expected count. The shot noise for a full-well capacity of 40,000 electrons is 200 electrons. The maximum signal-to-noise (SNR) ratio is 200. The shot noise occupies 8 bits (256). Thus, a 16-bit full-well capacity has an 8-bit range above the shot noise. An 8-bit digital video signal should be sufficient for a maximum contrast at 200 SNR with shot noise. The maximum contrast with shot noise occurs at the full-well capacity.

There are two common modes of frame transfer in CCDs for microscopes: full-frame transfer and interline transfer. A frame is defined by the pixel format, such as 1000 × 1000. A typical data rate is 40 MHz for an 8-μm pixel. The frame rate for this example is approximately 40 Hz. Additional operations per frame may reduce the frame rate to 30 Hz. The read noise increases with data rate due to a larger temporal bandwidth. The dark-current noise decreases with frame rate due to a longer temporal integration. The data and frame rates should be chosen wisely with regard to the application.

A full-frame transfer device requires a CCD array with two adjacent frames on the same chip. A bright frame is exposed to the image. A dark frame is buried beneath an opaque shield. A full frame is transferred from the bright frame to the dark frame in approximately 1000 data cycles. A data cycle is 0.025 ms at 40 MHz. A full-frame transfer to the video signal requires one million data cycles, which is 25 ms. Additional photoelectrons are collected in the bright frame during frame transfer to the dark frame. A bright spot during frame transfer can smear into a dark background.

An interline-frame transfer device requires an array of bright and dark columns. A bright column is exposed to the image during integration. A dark column is buried beneath an opaque shield. A bright column transfers charge to a dark column in one data cycle. The dark column is transferred to the video signal while the bright column collects more light. A progressive scan delivers an entire frame to the video signal in a sequence of contiguous rows during one million data cycles, or 25 ms. A progressive scan may employ dual outputs toward a double frame rate. An interlaced scan delivers half of the frame during 0.5 million data cycles, or 13 ms. However, there is a delay of 13 ms between the odd and even rows. Consequently, a rapidly moving feature may not correlate properly between odd and even rows.

An interline-transfer CDD might employ a microlens as shown in Fig. 15.3. The microlens is spherical or cylindrical. Its shape is determined by the surface tension of a polymer during solidification. A microlens collects light over the entire pixel. A spherical lens directs light from a lens stop into a spot on the full-frame sensor. A cylindrical microlens directs light from a lens stop into a bright column of an interline sensor. The microlens of Fig. 15.3 may increase the quantum efficiency of an interline sensor from 10 to 40%. The projected lens stop of the microlens is defined by a distant image of the bright column. Consequently, the microlens is telecentric. An F/2.8 microlens defines a half angle of collection at 10 deg. Many wide field lenses cannot accommodate this telecentric acceptance angle.

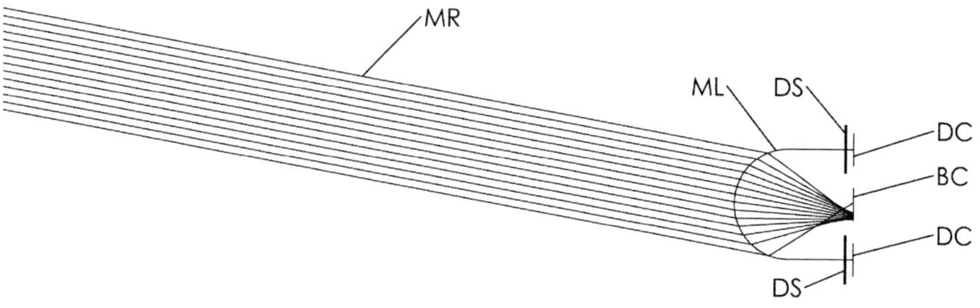

Figure 15.3 Microlens (ML) of a CCD. The microlens is centered over the bright column (BC). Dark columns (DC) are located beneath dark shields (DS). Marginal rays (MR) from the lens stop (not shown) are directed onto the bright column.

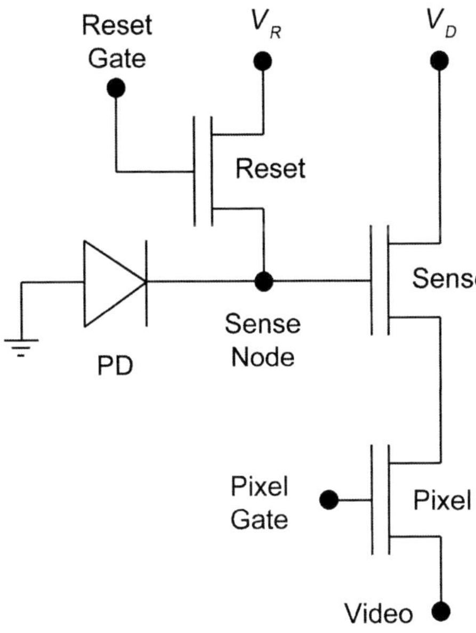

Figure 15.4 Circuitry of an active pixel sensor (APS). An APS contains a photodiode (PD) and three transistors. The signal is delivered directly to the video signal. Drive voltage V_D sends current through the sense FET and load resistor R_L. The reset gate restores the sense node to the reference voltage V_R.

15.2 Active Pixel Sensors

An active pixel sensor (APS) employs amplification of a photodiode output within each pixel. Figure 15.4 displays a three-transistor circuit for an APS. Each pixel contains a photodiode, sense transistor, select transistor, and reset transistor. An APS frequently employs the same fabrication methods as CMOS technology for flash drives, which can tolerate large amounts of noise. A CMOS APS normally displays much more noise than a similar CCD sensor.

A Kodak CMOS sensor KAC-00401 provides an example of CMOS noise at 20° C. Its pixel dimension is 6.7 μm. The typical dark current is 0.9 nA/cm^2, which translates to 2100 electrons per pixel per second. At 30 frames per second (fps), the dark current contributes 70 photoelectrons. The typical dark current is 39 mV per pixel per second. At 30 fps, the dark current contributes 1.3 mV. There is ~ 1 V per 50,000 electrons at the sense node. The typical read noise is 4.3 bits within a 12-bit analog-to-digital converter. If 41,000 photons are scaled to a digital count of 4096 (12 bits), then 1 bit equals 10 photons, and a 4.3-bit read noise equals 200 photons. The combination of dark current and read noise yields a total dark noise of 270 photoelectrons at 30 fps. This is nine times greater than the CCD sensor. After consideration of the smaller area of the CMOS pixel, the noise per area of the CMOS device is 10 times greater than that of the CCD device. These specifications are readily available from the Kodak device specifications for the KAC-00401.[36]

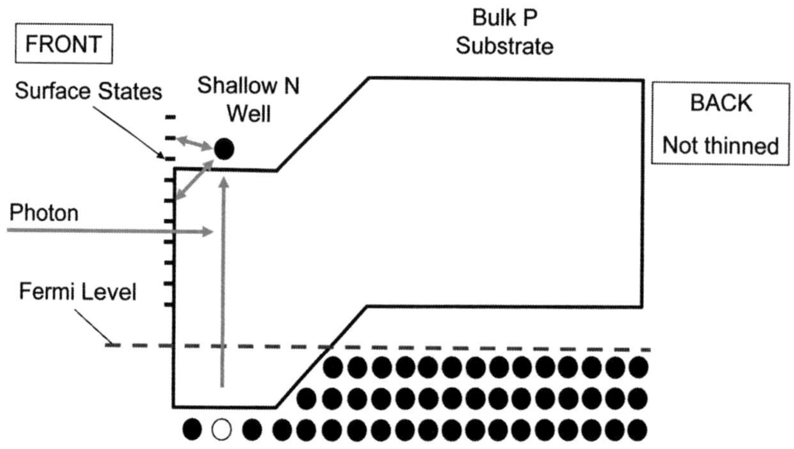

Figure 15.5 Potential well of a CMOS detector. Photon absorption promotes electrons (filled circles) from valence band to conduction band. N-implantation of the bulk p-substrate creates a potential well for photoelectrons. The shallow well is exposed to surface states. CMOS foundries do not typically provide deep n-implantation or shallow p-implantation.

Figure 15.5 displays a typical structure of a CMOS detector. The n-implantation forms a shallow well with exposure to surface states. Many CMOS foundries do not offer deep n-implantation or shallow p-implantation. These processes are not required for flash memory. Most CMOS devices operate at 3.3 V, which offers lower power consumption than a CCD at 5 volts.

15.3 Photomultiplier Tubes

A photomultiplier tube (PMT) defines the vision field as a single pixel. It is reasonable to assign a pixel collection efficiency of 100% to a PMT, because the spot will likely fall completely within the single pixel. The dark current is extremely low. A high voltage of nearly 1000 V is required for operation. Its temporal response is limited by a resistor-capacitor (RC) circuit.

Figure 15.6 displays the electrical circuit of a PMT. The photocathode of the PMT has a low work function for electrons. Consequently, UV and visible photons may eject electrons into the vacuum of the PMT. The ejected electron is driven by an electric field toward a first dynode. The impact of the electron on the dynode ejects multiple electrons, which are driven toward a second dynode. This process repeats at numerous dynodes until termination on the anode capacitor and load resistor.

A PMT may operate in reflection or transmission mode. A reflection-mode PMT employs a reflective photocathode, as illustrated in Fig. 15.6. The electrons are ejected from the front of the photocathode as if they were reflected. A transmission-mode PMT employs a transmissive photocathode. The electrons are ejected from the photocathode as if they were transmitted through the photocathode.

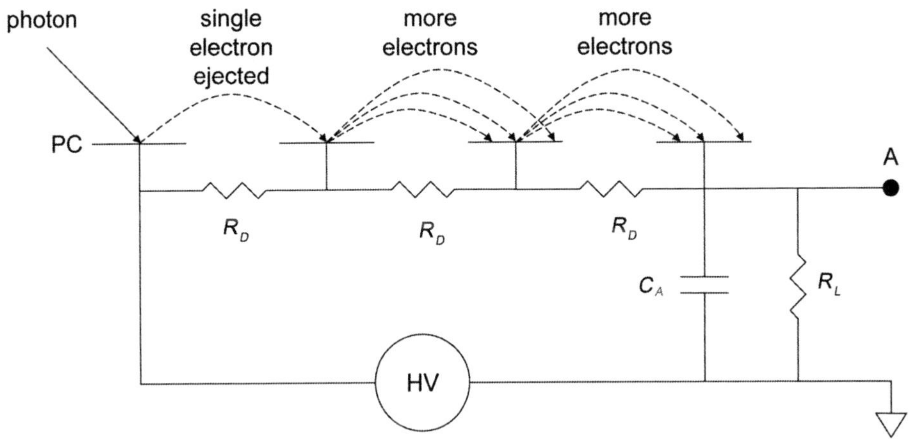

Figure 15.6 PMT circuit with components: photocathode (PC), dynode (D), dynode resistor (R_D), anode (A), anode capacitor (C_A), load resistor (R_L), and high voltage (HV) (500–800 V).

The charge accumulation by the anode capacitor is expressed as

$$Q = \Phi_I \eta_Q g_D \tau_I, \tag{15.1}$$

where Φ is the incident flux on the detector, η_Q is the quantum efficiency of conversion of photons into electrons, g_{DG} is the efficiency of the gain in the dynode structure, and τ_I is the time of integration.

The temporal decay of the output capacitor is described as

$$V = \frac{Q}{C_A} \exp\left(\frac{-t}{R_L C_A}\right), \tag{15.2}$$

where C_A is the anode capacitance and R_L is the load resistance across the anode capacitor. A second load resistor may be added in parallel to the first for increased speed by addition of a second discharge path.

A reflection-mode PMT requires a transparent widow for passage of photons into the vacuum and on the reflection-mode photocathode. The electrons are excited near the surface where a strong electric field drives them away from the photocathode toward the first dynode. There is little chance for an electron to relax to a bound state within a reflection-mode PMT. The spatial response of a reflection-mode photocathode is very nonuniform. There is a strip of maximum quantum efficiency at the side of the cathode in proximity to the first dynode.

A transmission photocathode is a thin film on the inside of silica window. The photoelectrons must travel through the film before entering the vacuum. Consequently, there is significant opportunity for relaxation to bound states within the photocathode. The quantum efficiency is compromised by the transmission mode. However, the quantum efficiency is spatially uniform across the transmission-mode photocathode. Such uniformity promotes consistency between instruments.

15.4 Film

Photographic film employs grains of silver halide (AgBr, AgCl, AgI), which convert photons into metallic silver. This process has three main steps. First a photon is converted into a valence electron with mobility in the grain. Second, an excited electron falls into a trap as an invisible record of photon absorption. Third, a developer converts the excited silver halide into colloidal metallic silver.

Silver displays one of the largest intrinsic reflectance of any metal within the visible wavelength regime. An interpolation between 2.0 and 2.5 eV estimates the refractive index as $0.263 + i3.94$ at the d line (2.11 eV). The magnitude of the complex vector is 3.95. Within air at index of 1.00, this yields a reflectance of 94% at a maximum absorbance of 6%. Within an emulsion at an index of 1.50, this yields a reflectance of 92% at a maximum absorbance of 8%. Most of the optical density is due to reflectance. Consequently, the generation of heat is minimized.

Figure 15.7 displays an exposure curve for a monochrome film for visible photography. Table 15.1 displays specific parameters. The 400-speed film has a 3.7-μm grain, while the 100-speed film has a 7.5-μm grain. The 400-speed film responds to a lower exposure as measured in fluence (Talbots per area).

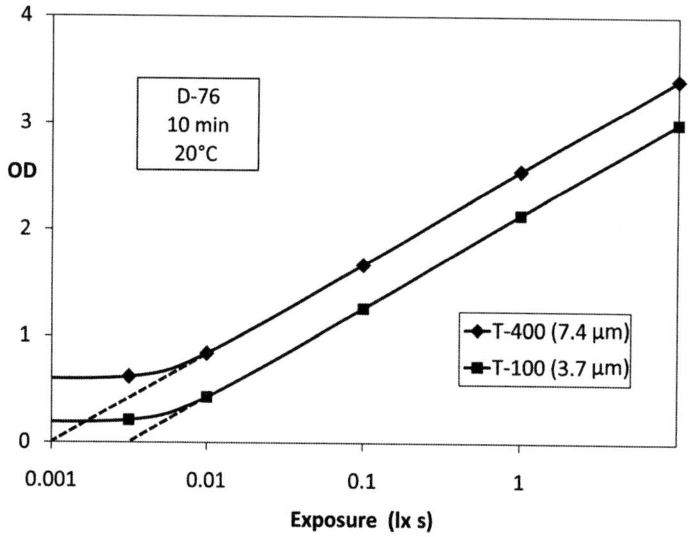

Figure 15.7 Optical density of photographic film, exhibiting exposure curves of monochrome film at speeds of 400 and 100. Exposure is measured in fluence [Talbots per area (lx·s)]. Optical density was defined in Eq.(13.2). Parameters of T-400: 60 cyc/mm as a high-frequency limit, 7.5-μm Gaussian pixel, 0.004 lx·s. Parameters of T-100: 120 cyc/mm as high-frequency limit, 3.7-μm Gaussian pixel, 0.001 lx·s. The threshold count is 230 photons for both T-100 and T-400. Both require development in D-76 for 10 min at 20° C. Data points were measured from plots within the Kodak specification F-4016.

Table 15.1 Monochrome film parameters.

	Frequency (cyc/mm)		Lens stop		Pixel size	Read noise		
Film	**0.5 MTF**	**0.0 MTF**	**NA_G**	**NA_M**	**(μm)**	**(lux·s)**	**($p/\mu m^2$)**	**(photons)**
T-400	60	120	0.10	0.14	7.5	1×10^{-3}	4	230
T-100	120	240	0.20	0.28	3.7	4×10^{-3}	16	230

However, the 400-speed film has a larger dark threshold. Both films exhibit the same dark threshold at 230 photons per crystal. The 100-speed film exhibits better contrast with shot noise because 4X as many photons are required for the same optical density. At a 4X photon count, the contrast with shot noise is 2X greater [Eq. (8.6)]. In terms of decibels, at a specific OD, the T-100 film requires 6 dB more photons toward a 3-dB greater contrast with shot noise.

The marginal NA of the 400-speed film nearly matches that of a typical tube lens, 0.025. Consequently, the 400-speed film is optimized for microscopy. However, as mentioned earlier, the larger pixel of the 100-speed films creates better contrast with shot noise.

Frequency data are measured from modulation transfer curves of the Kodak specifications for T-MAX film.[37] The marginal NA is derived from the cutoff frequency in Eq. (19.37), and the Gaussian NA is derived from Eq. (5.49).

Chapter 16
Human Vision

16.1 Physiology

There are three basic components of human vision: the point spread of the lens, the lateral inhibition of the retina, and the saccation of the eye. Each component is described by a MTF, which acts on the irradiance. The physiology of the retina includes rods, cones, horizontal cells, bipolar cells, amacrine cells, and ganglion cells.[38]

The human eye evolved in numerous steps.[39] The eye began as a simple photoreceptor that detected light and shadow. Then, an eye cup created a camera[i] with a finite angular field of view. A retina created an image. A fixed lens sharpened the image. A variable iris controlled brightness. And lastly, a dynamic lens controlled focus. Even Charles Darwin confessed to such a progression by natural selection as "absurd."[ii] However, Darwin did believe that incremental improvements could define a progression by natural selection. The intermediate steps are present within vertebrates of the current day.[39]

The fovea[iii] comprises tightly packed cones with very few rods. The absence of blood vessels and other cells creates a pit that defines the spatial limit of the fovea. The foveola (the center of the fovea) comprises red, green, and blue cones without any rods. The annular portion beyond the foveola contains a few rods but is still largely clear of blood vessels, bipolar cells, amacrine cells, and ganglion. The current model employs an average cone diameter at 3.8 um. At this diameter, in a hexagonal packing scheme, the foveola contains 5000 cones within a 200-μm diameter, while the fovea contains 20,000 photoreceptors within a 400-μm diameter.[40] Within the fovea, there is a single incremental ganglion per photoreceptor. Beyond the fovea, rods are more common.

The photoreceptors connect to bipolar cells that feed into ganglia. An incremental bipolar-to-ganglion path responds to an increment of light. Sjostrand et al.[40] calculated three ganglia per foveal cone from experimental quantification of both. Thus, a single cone within the fovea connects to three ganglia: one incremental ganglion, one decremental ganglion, and a third type, such as a parasol ganglion. An incremental bipolar-to-ganglion path responds to an increment of light. A decremental bipolar-to-ganglion path responds to a

[i] A camera is a closed chamber with a hole. From Greek *kamera*, meaning vault.

[ii] See full quotation in Lamb *et al.*[39]

[iii] *Fovea*, Latin, meaning pit.

decrement of light. The bipolar and ganglion cells of the fovea are laterally displaced beyond the fovea. The displacement of these cell bodies creates a pit.

Horizontal cells create lateral inhibition from adjacent photoreceptors. The horizontal cells secrete gamma amino butyric acid (GABA), which inhibits transduction from the photoreceptors to the bipolar cells.

The periphery is defined by the region beyond the fovea, where blood vessels and other cells are prevalent. These features of the peripheral retina may scatter incident light, which degrades the optical point spread. Both cones and horizontal cells grow larger with distance from the fovea. A receptor pool comprises multiple receptors with connection to a single ganglion. The effective size of a peripheral photoreceptor is much larger than a foveal receptor, due to receptor pooling. The morphology of a peripheral horizontal cell is much different from a foveal horizontal cell. Numerous features of the periphery degrade spatial resolution as defined by a point spread of the system.

This model is a practical implementation of transfer functions of the lens, retina, and saccation. The variety of cone diameters is represented by a Gaussian point spread. A foveal cone is connected to a single incremental bipolar-to-ganglion path.

16.2 Contrast Sensitivity Function

Contrast sensitivity is defined by the reciprocal of the threshold visibility. The peak contrast sensitivity of human vision is approximately 160 at 275 cyc/rad for the natural pupil at a 2.5-mm diameter. The high-frequency limit occurs at about 2500 cyc/rad at a contrast sensitivity of 1.

The contrast sensitivity function (CSF) is modeled by the product of transfer functions of the lens, retina, and saccation:

$$\mathrm{CSF}(k) = 160 \frac{G_L(k) G_R(k) G_S(k)}{G_L(k_P) G_R(k_P) G_S(k_P)}, \tag{16.1}$$

where k is the spatial frequency, $G_L(k)$ is the transform of the lens, $G_R(k)$ is the transform of the retina, $G_S(k)$ is the transform of saccation, and k_P is the spatial frequency of the peak contrast sensitivity. Each transform is described in detail in the subsequent sections.

Figure 16.1 displays the CSFs of three pupil diameters. Each pupil requires a different radius of saccation for maximum contrast with the lens-point point spread. Obviously, the artificial pupil displays higher peak frequency than the natural pupil. Consequently, the artificial pupil creates a sharper image. The exit pupil of a microscope is an extremely important part of the optical system.

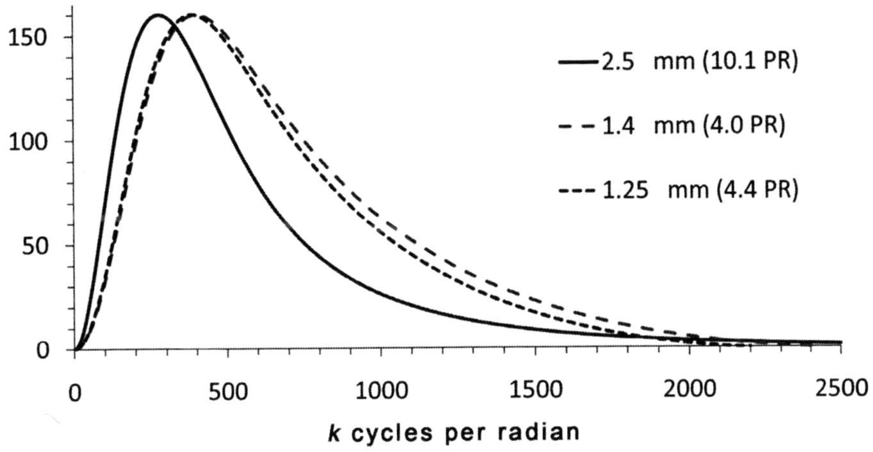

Figure 16.1 Contrast sensitivity plot of human vision. The angular diameter of a photoreceptor (PR) is 0.23 mrad. The angular diameter of a horizontal cell is 13 times that of the PR at 3.0 mrad. Each plot is identified by a pupil diameter and a saccation radius. The natural pupil diameter at 2.5 mm employs a saccation radius at 6.8 PR diameters toward a peak sensitivity at 280 cyc/rad. The artificial pupil diameter at 1.4 mm employs a saccation radius at 2.7 PR diameters toward a peak sensitivity at 400 cyc/rad. The artificial pupil diameter at 1.25 mm employs a saccation radius at 3.0 PR diameters toward a peak sensitivity at 385 cyc/rad.

16.3 Point Spread of a Lens

There are two important formats of the point spread of a lens: a diffraction-limited format of an artificial pupil and an aberration-limited format of the natural pupil. The exit pupil of a microscope frequently defines an artificial pupil at a 1.4 or 1.25 mm diameter. The natural pupil of the eye is defined as 2.5 mm, which is typical for an office or a laboratory environment. The effective focal length of the eye is 16.7 mm. Figure 16.2 displays the MTFs of the lens at three pupil diameters.

The transform of an artificial pupil at 1.4 mm is defined by the diffraction limit of the pupil [Eq. (19.32)]. After Eq. (19.33), the cutoff frequency for an artificial pupil is

$$k_{AP} = \left(\frac{2\mathrm{NA}_{AP}}{\lambda}\right) f_E \ (\text{cycles per radian}), \tag{16.2}$$

where NA_{AP} is the NA of the artificial pupil and f_E is the focal length of the eye. The artificial NA of a 1.4-mm pupil is 0.042, which indicates a cutoff frequency of 2444 cyc/rad for the d line. The angular extent of an Airy disk of this lens is derived from Eqs. (5.19) and (1.2) as

$$\beta_{Airy} = \frac{\phi_{Airy}}{f_E} = 2.44\frac{\lambda}{\phi_{AP}}, \tag{16.3}$$

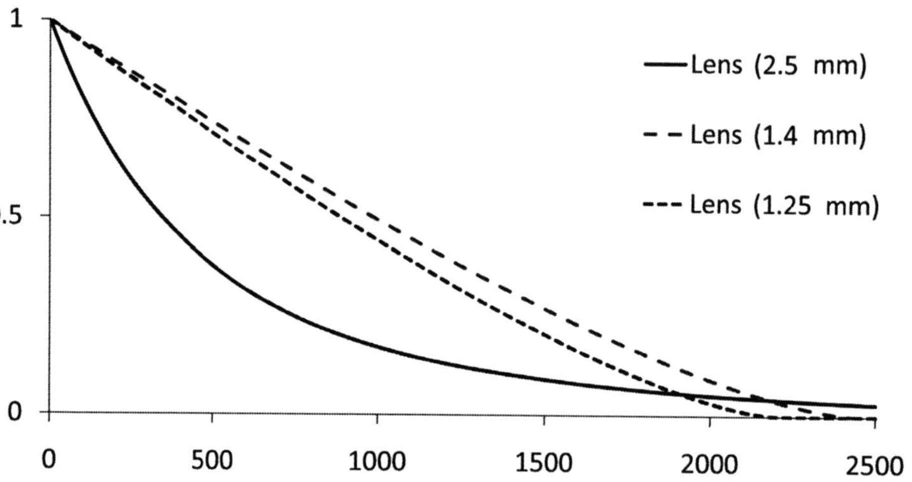

Figure 16.2 Transformations by a lens of human vision. The 2.5-mm pupil is based on the model of Artal and Navarro.[41] The 1.4-mm pupil is based on the Airy pattern of an exit pupil of a 180-mm tube lens. The 1.25-mm pupil is based on the Airy pattern of an exit pupil of a 200-mm tube lens.

where ϕ_{AP} is the diameter of the artificial pupil. The angular extent of the Airy disk is 1.0 mrad (3.5 arcmin) at 1.4 mm and 587 nm. The equivalent Gaussian diameter is 0.7 mrad.

The transform of an artificial pupil at 1.25 mm defines an 0.037 artificial NA and a cutoff frequency of 2200 cyc/rad for the d line. The angular extent of the Airy disk is 1.15 mrad (3.9 arcmin) at 1.25 mm and 587 nm. The equivalent Gaussian diameter is 0.8 mrad.

The transform of the lens at the natural 2.5-mm pupil is accurately described by Artal and Navarro.[41] Their empirical transform of the natural pupil $G_{NP}(k)$ acts on flux per angle:

$$G_{NP}(k) = (1 - C_{NP})\exp(-A_{NP}k) + C_{NP}\exp(-B_{NP}k), \tag{16.4}$$

where k is an angular frequency, and the constants of the natural pupil are defined as follows:

$$A_{NP} = 2.8 \text{ mrad/cyc}, \quad B_{NP} = 1.0 \text{ mrad/cyc}, \quad C_{NP} = 0.36. \tag{16.5}$$

This relation is valid up to 50 cycles per degree, or 2900 cyc/rad.

16.4 Lateral Inhibition of the Retina

The lateral inhibition of the retina is created by the photoreceptors, the horizontal cells, and the bipolar cells. Figure 16.3 displays a simple model of transduction in

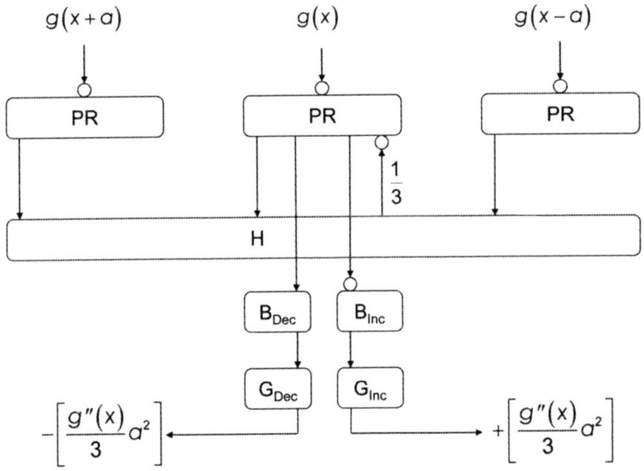

Figure 16.3 Transduction circuit of the retina. Each node represents an electric potential of a cell. There are two feed-forward paths. The incremental path comprises: a photoreceptor (PR), an incremental bipolar cell (B_{Inc}) with an inverted input, and an incremental ganglion (G_{Inc}). The decremental path comprises: a photoreceptor, a decremental bipolar cell (B_{Dec}), and a decremental ganglion (G_{Dec}). The inverted feedback of the horizontal cell (H) on the center photoreceptor provides lateral inhibition from the surrounding photoreceptors.

the retina. Upon exposure to light, a photoreceptor acquires more negative charge; thus, its electric potential decreases in response to an increment of light. There are two feed-forward paths: the incremental path creates a positive response to an increment of light, and the decremental path creates a positive response to a decrement of light.[iv] Tessier-Lavigne[42] provides an effective introduction to the transduction of the retina with simple figures. Sterling[43] provides a more rigorous discussion through numerous citations of empirical data.

The negative feedback of the horizontal cells creates lateral inhibition of on-center bipolar cells from off-center photoreceptors. This defines a condition of center-surround antagonism. The input signals to the photoreceptors may be expressed by a Taylor series:[6]

$$\begin{aligned} g(x-a) &= g(x) - \frac{g'(x)}{1!}a + \frac{g''(x)}{2!}a^2 \\ g(x+a) &= g(x) + \frac{g'(x)}{1!}a + \frac{g''(x)}{2!}a^2. \end{aligned} \tag{16.6}$$

[iv] In the field of neuroscience, the incremental path is the "on-center" path, while the decremental path is the "off-center" path. A negative electric potential of a cell is called "polarization." *Hyper*polarization *increases* the magnitude of the negative electric potential across the cell membrane. *Hypo*polarization *decreases* the magnitude of the negative electric potential across the cell membrane.

The summation of signals at the incremental ganglion is expressed as

$$+g(x)-\left[\frac{g(x-a)+g(x)+g(x+a)}{3}\right]. \tag{16.7}$$

Combining Eqs. (16.6) and (16.7) reveals a line spread of the retina as

$$g_R(x)=\frac{1}{3}g''(x)a^2. \tag{16.8}$$

The dependency on the second derivative indicates the retina as a detector of spatial curvature as opposed to magnitude or slope. This dependency on spatial curvature is very close to its actual function. Human vision is based on curvature and not on magnitude.

The line spread of the fovea may be also described by a difference of Gaussians:

$$g_F(x)=\exp\left(\frac{-4x^2}{\beta_{PR}^2}\right)-\exp\left(\frac{-4x^2}{\beta_H^2}\right). \tag{16.9}$$

where β_{PR} is the full angle of the photoreceptor and β_H is the full angle of the horizontal cell. This equation effectively represents the receptor fields within the fovea, but not the periphery. The transfer function of the fovea is

$$G_F(k)=\exp\left(\frac{-\pi^2\beta_{PR}^2k^2}{8}\right)-\exp\left(\frac{-\pi^2\beta_H^2k^2}{8}\right), \tag{16.10}$$

Figure 16.4 displays the transform of the fovea.

16.5 Temporal Feedback of Photoreceptors

There is also temporal feedback within a photoreceptor [44,45] that drives the gain of the photoreceptor to zero in response to a steady input of light. The peak of temporal contrast sensitivity occurs at approximately 15 Hz for an illuminance of 200 lux, which is typical for office lighting. The cutoff frequency for office lighting is approximately 80 Hz. According to the data of Kelly,[44,45] both of these frequencies increase as illuminance does.

The temporal feedback is countered by saccation.[46] A saccade defines a motion within the limits of a small volume or “sack.” A large saccade of > 10 photoreceptors occurs on average every 200 ms. A microsaccade of < 10 photoreceptors occurs every 12 ms, which specifies a rate of 80 Hz. The microsaccades are essential for continuous regeneration of the image in the presence of temporal feedback.

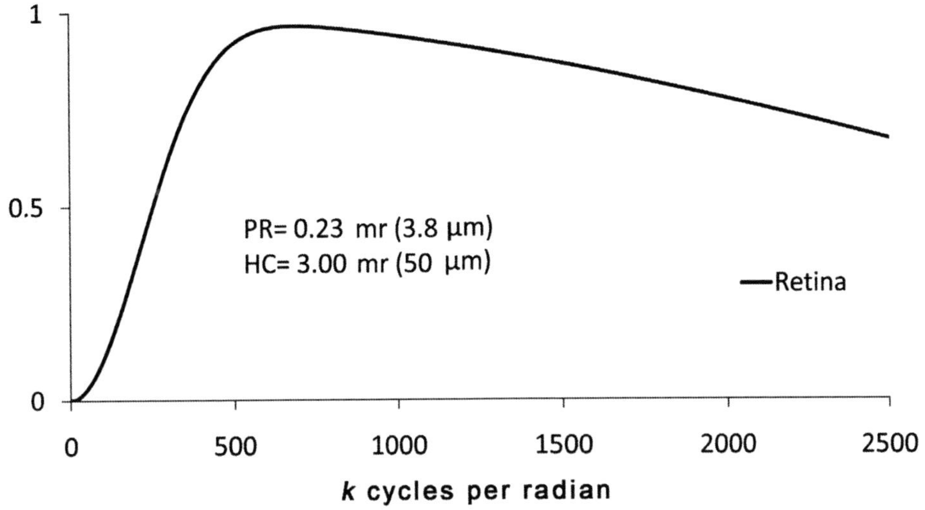

Figure 16.4 Transformation of the fovea of human vision. A photoreceptor is approximately 0.23 mrad (0.8 arcmin, 3.8 μm) in Gaussian diameter. A horizontal cell is 13 times larger at approximately 3.0 mrad (10 arcmin, 50 μm).

16.6 Saccation Point Spread

The probability of saccation within a radius is an exponential function as derived by Steinmann:[46]

$$P_S(r) = 1 - \exp\left(\frac{-r}{\alpha_S}\right), \tag{16.11}$$

where α_S is the mean of saccation. The point spread of saccation is

$$g_S(r) = \exp\left(\frac{-r}{\alpha_S}\right). \tag{16.12}$$

The line spread of saccation is described as

$$g_S(x) = \exp\left(\frac{-x}{0.6\alpha_S}\right). \tag{16.13}$$

The transfer function of saccation is derived from the line spread as

$$G_S(k) = \frac{1}{1 + i2\pi k(0.6\alpha_S)}. \tag{16.14}$$

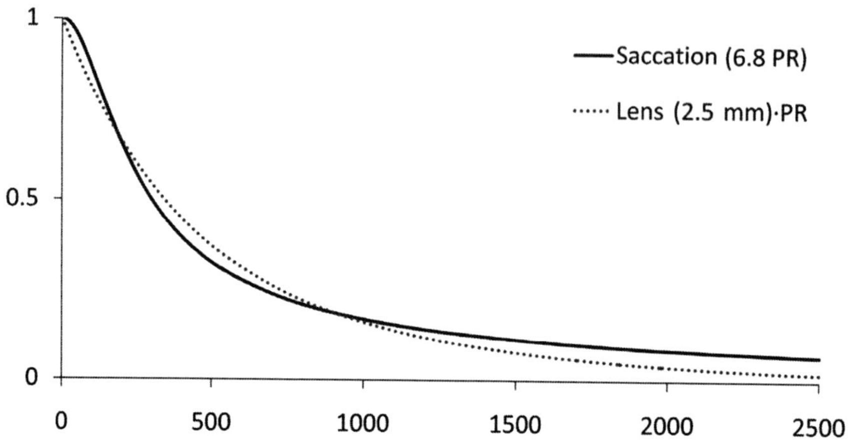

Figure 16.5 Correlation of saccation transform to other transforms. Product of transforms of the lens and photoreceptor [Lens (2.5 mm)·PR] establishes a target for the saccation transform with a radius of 6.8 PR diameters.

As shown in Fig. 16.5, the magnitude of a saccation transform is matched to the product of transforms of the lens and photoreceptor. The lens point spread is typically much larger than the photoreceptor. If the output signal of the retina controls the magnitude of saccation, then the retina should drive the saccation magnitude toward the point spread of the lens and photoreceptor. Consequently, the saccation angle is adjusted for maximum correlation between the saccation transform and the combined transform of the lens and photoreceptor.

Figure 16.6 displays the saccation transforms in optimum correlation to the lens transforms at three common pupil diameters. The smaller pupils are artificial pupils. The exit pupil of a microscope or telescope may significantly sharpen the point spread of the eye as indicated by the larger MTF at higher frequencies.

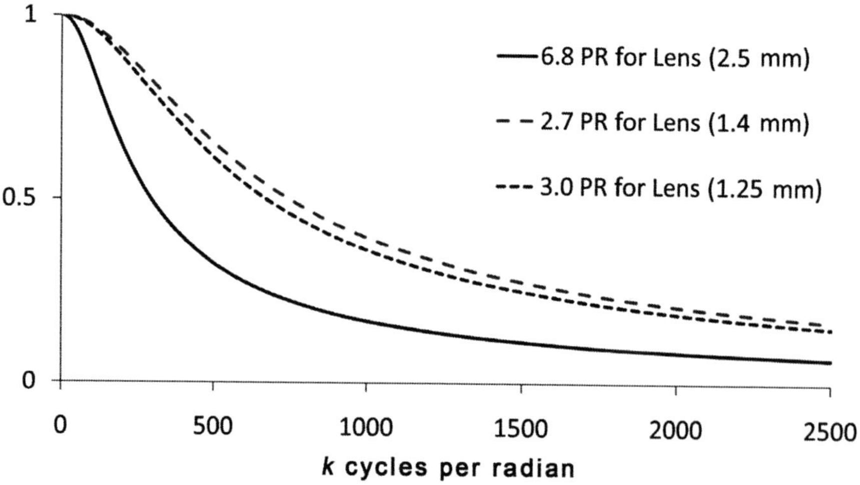

Figure 16.6 Transformation by saccation of human vision. Saccation radius versus pupil diameter.

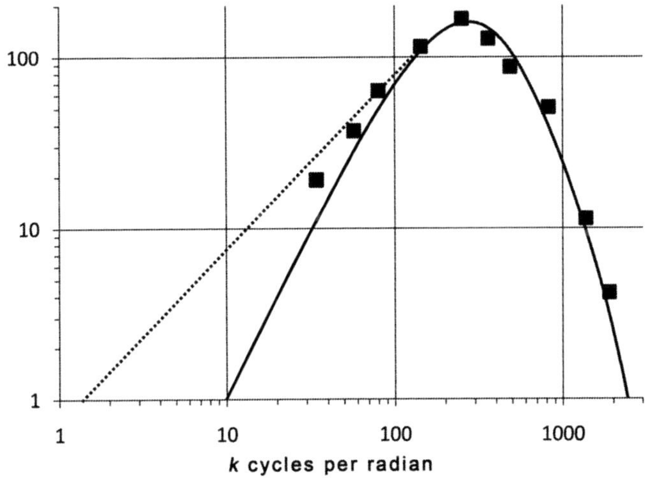

Figure 16.7 Contrast-sensitivity plot of human vision in log-log format. The parameters of the plot are the same as the natural pupil of Fig. 17.1. Peak sensitivity is 160 at 275 cyc/rad for the natural pupil at 2.5 mm. The experimental data is an average of five subjects. The low-frequency slope of 2 indicates a frequency-squared dependence, which is a departure from reality. At low frequencies, the experimental slope approaches 1. Data points were derived from application of a CAD program to a scanned image of De Valois.[47] Theoretical data is plotted at increments of 5 cyc/rad (5, 10, 15... cyc/rad).

16.7 Vision Research

Vision research indicates the study of the retina and the brain. The shape of a CSF plot in log-log format reveals important information about transduction. In the low-frequency portion, a frequency-squared dependency indicates a Gaussian field of a horizontal cell. Conversely, a linear frequency dependency indicates an exponential decay as the field of a horizontal cell. These low-frequency dependencies are certainly relevant to vision research; however, they are not as important for instrument design.

Figure 16.7 displays a log-log plot of the CSF with experiment data from DeValois.[47] The peak spatial frequency is near 250 cyc/rad. Thus, a 2 mrad object matches the central lobe of the point spread of the natural pupil. The high-frequency limit is near 2500 cyc/rad. The low-frequency limit is < 10 cyc/rad.

Peripheral spatial vision requires numerous modifications to the model. The photoreceptors become larger with distance from the fovea.[48] They are also directed into pools by lateral connections. A horizontal cell in the periphery has fewer and longer protrusions.[49] A peripheral horizontal cell should be modeled as an exponential decay. Such dependencies are important to neural science.

16.8 Temporal Contrast Sensitivity Function

The temporal CSF displays two components: a feed-forward and a feedback. A feed-forward transduction may resemble the charging of a resistor and capacitor (RC). The transform of an RC circuit is

$$G_{RC}(\omega) = \left| \frac{1}{1 + i\omega\tau_{RC}} \right|, \qquad (16.15)$$

where ω is the temporal frequency and τ_{RC} is the time constant of the RC circuit. A cascade of four similar reactions creates a feed-forward transform

$$G_{FF}(\omega) = \left| \frac{1}{1 + i\omega\tau_{FF}} \right|^4 = \left[\frac{1}{1 + (\omega\tau_{FF})^2} \right]^2, \qquad (16.16)$$

where τ_{FF} is the feed-forward time constant. The exponent of 4 represents a cascade of four similar feed-forward reactions. The feed-forward process is governed by electric transduction of ions.

The low-frequency limit of the temporal CSF is estimated by the feedback of the horizontal cells. A cascade of two similar reactions creates a feedback transform

$$G_{FB}(\omega) = \left| 1 - \exp(-i\omega\tau_{FB}) \right|^2 = 1 - \cos(\omega\tau_{FB}), \qquad (16.17)$$

where τ_{FB} is the feedback time constant. The feedback path contains two nodes: the horizontal cell and photoreceptor-to-bipolar synapse. The feedback process is governed by secretion of GABA from the horizonal cell.[43] GABA opens anion channels of the photoreceptor.

Figure 16.8 displays the feed-forward and feedback components of the temporal CSF. The temporal CSF is effectively modeled as

$$\mathrm{CSF}(\omega) = 135 \frac{G_{PR}(\omega) G_{HC}(\omega)}{G_{PR}(\omega_P) G_{HC}(\omega_P)}, \qquad (16.18)$$

where ω_P is the frequency of peak contrast sensitivity at 135. The horizontal cell feedback defines the low-frequency response, while the photoreceptor feed forward largely defines the high-frequency response.

Figure 16.9 displays the temporal CSF of the human vision model in this text and the experimental data of Kelly.[44] The lighting for the experimental data at 174 cd/m^2 is similar to a computer monitor, which is normally 50–300 cd/m^2. The peak contrast sensitivity at 135 occurs at a temporal frequency of 15 Hz. The high-frequency limit is near 80 Hz.

The temporal CSF model certainly makes a few broad assumptions. However, the theoretical model does resemble the experimental data. Ergo, it has practical value for instrument design.

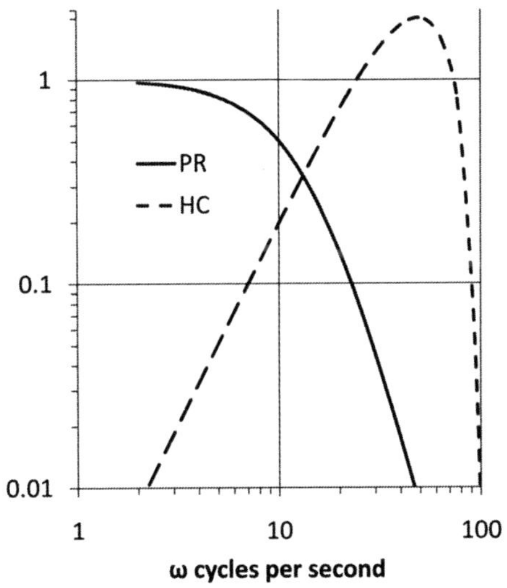

Figure 16.8 Components of a temporal contrast sensitivity plot. Photoreceptors (PR) define the high-frequency dependency. Horizontal cells (HC) define the low-frequency dependency. HCs provide maximum feedback at 50 Hz.

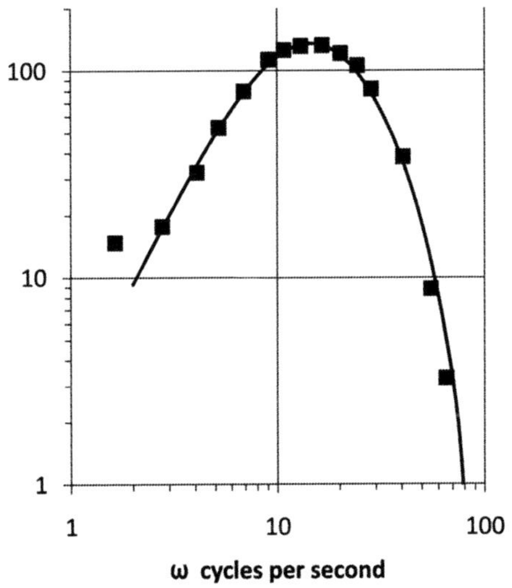

Figure 16.9 Temporal CSF. Peak sensitivity at 135 occurs at 15 cyc/rad for the natural pupil at 2.5 mm. The feed-forward and feedback time constants are identical at 10.2 ms. The pupil has a natural diameter of 2.5 mm. Luminance is 175 cd/m^2. Experimental data points were derived from application of the CAD program to a scanned image of Kelly.[44] The theoretical plot occurs at multiples of 1.2 (2.0, 2.4...) cyc/rad.

Chapter 17
Optical Materials

17.1 Glass Types

The term glass implies an amorphous structure. Crystallization is defeated by either rapid cooling or diversity of components. Quartz is a crystalline form of silicon dioxide that is formed under intense pressure and heat. Silica is an amorphous form of silicon dioxide that is formed during rapid solidification.

Glass is largely composed of the nine most common elements within the earth's crust (Table 17.1). Additional components of glass are listed in Table 17.2. Iron is not desirable for optical glass due to its green color. Arsenic (As) and antimony (Sb) are fining agents that remove bubbles and impurities such as iron. Figure 17.1 displays a periodic table of common elements in metal-oxide glass.

Silica provides the bulk material for most optical glass. The liquidus temperature of silica is 1715° C, and the melting point is 2000° C. The liquidus temperature defines the transition from crystalline to amorphous structure. The liquidus temperature is also called the glass temperature T_g, above which the structure is amorphous. The melting point defines the transition from solid to liquid.

Soda lime is an early form of glass. The liquidus temperature of silica is reduced to 1000° C by the addition of soda Na_2O. The viscosity is reduced by addition of lime (CaO). Potassium oxide (K_2O) may replace soda. A low liquidus temperature and low viscosity are beneficial for blowing, shaping, and floating. A trace amount of iron oxide content creates a bluish-green tint, which is prominent at the edges of soda lime glass. A large amount of iron-oxide content creates brown bottle glass.

Crown glass was originally formed by a rod with molten glass on the end. The linear momentum of the rotation stretched the molten glass into a disk, but not evenly. The process left a crown in the middle. A high transmittance for windows was the primary goal in crown glass.

The original flint glass was comprised of silica and flint. Flint occurs naturally as quartz with trace elements. The essential trace elements raise the refractive index of silica without the addition of color. Naturally occurring flint provides sparkle without color.

Table 17.1 Abundant elements of glass in the earth's crust by weight.

Rank	Sym	Fraction	Formula	gm/cm^3	Name	Purpose	Natural forms Name	Natural forms Formula
1	O	0.474						
2	Si	0.277	SiO_2	2.6	Silica	Base network	Quartz	SiO_2
3	Al	0.082	Al_2O_3	4.0	Alumina	Stability	Sapphire	Al_2O_3
4	Fe	0.041	Fe_2O_3	5.2	Hematite, iron (III) oxide	Trace element with color	Rouge	Fe_2O_3
5	Ca	0.041	CaO	3.4	Lime	Lower viscosity	Calcite	$CaCO_3$
6	Na	0.023	Na_2O	2.3	Sodium oxide	Lower T glass	Soda ash	Na_2CO_3
7	K	0.021	K_2O	2.4	Potassium oxide	Lower T glass	Pot ash	K_2CO_3
8	Mg	0.023	MgO	3.6	Magnesia	Trace element	Dolomite	$MgCa(CO_3)_2$
9	Ti	0.006	TiO_2	4.2	Titania, titanium (IV) oxide	Electron density	Rutile	TiO_2
Total		0.988						

Data from webmineral.com and webelements.com.

Table 17.2 Rare elements of glass in the earth's crust by weight.

Sym	ppm	Formula	gm/cm^3	Name	Purpose	Natural forms Name	Natural forms Formula
As	2	As_2O_3	4.3	Arsenic oxide	Removes iron	Arsenopyrite	FeAsS
B	950	B_2O_3	2.5	Boron oxide	Lowers TCE	Borax	$Na_2B_4O_7$:$10H_2O$
Ba	500	BaO	5.7	Barium oxide	Electron density	Barite	$BaSO_4$
Nb	20	Nb_2O_5	4.6	Niobium oxide	Electron density	Niobite	$Fe^{2+}Nb_2O_6$
Pb	14	PbO	9.6	Lead oxide	Electron density	Massicot	PbO
Sb	0.2	Sb_2O_3	5.2	Antimony oxide	Removes iron	Berthierite	$FeSb_2S_4$
La	32	LaO	6.5	Lanthanum oxide	Electron density	Bastnasite	(Ce, La)CO_3F

Data from webmineral.com and webelements.com.

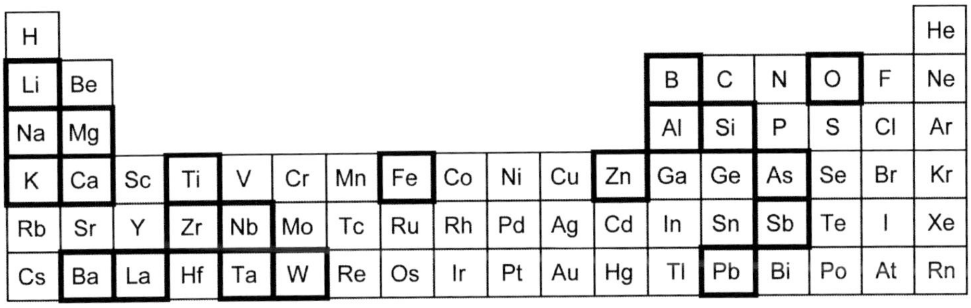

Figure 17.1 Periodic table of metal-oxide glass. Thick borders indicate common elements. Arsenic (As) and antimony (Sb) bind to iron (Fe) in the process of removal.

Lead glass eventually replaced flint glass. The addition of lead oxide raises the refractive index and lowers the liquidus temperature. A higher refractive index creates more sparkle. A lower liquidus temperature extends the temperature range for glass blowing and shaping. Lead is a common heavy metal because three out of four heavy nuclei decay into a stable form of lead. Lead glass employs arsenic as a fining agent that removes bubbles and iron.

Ecologically friendly glass is called ecoglass. Ecoglass is considered safe for the environment due to the absence of lead and arsenic. Titanium oxide has replaced lead oxide. Antimony has replaced arsenic as the fining agent in a similar manner to arsenic. Antimony bonds with iron and reduces viscosity, thus contrubuting to removal of bubbles. However, antimony is also similar in toxicity to arsenic. Fortunately, antimony is les soluble than arsenic. Arsenic is soluble in water as an arsenite anion, which is a significant problem in some water supplies.

The shift from lead to titanium was mandated by a directive of the European Community in 2002. The mandate forbid the sale of electrical products with lead, mercury, and hexavalent chromium from 2006 onward. The production of lead glass is still legal, while the sale of an electrical product with lead requires an exception to the law.

Float glass is derived from pulling molten soda lime glass across a bath of molten tin. The temperature of the bath is graded from a high to a low temperature. A solid glass ribbon is pulled off the low temperature end by rollers. Most plate glass for windows is a form of soda lime float glass with its characteristic green tint at the edges. Float glass is not appropriate for most optical instruments due to its color and surface form error.

Commercial borosilicate comprises mostly silica (90%) with small amounts of boron oxide (3%), sodium oxide (2%), aluminum oxide (2%), and potassium oxide (1%). Boron is a small atom that bridges dangling bonds within amorphous glass. Consequently, the coefficient of thermal expansion (CTE) is reduced by the boron. Glass cookware and labware such as Pyrex® and DURAN® is borosilicate.

Optical borosilicate crown (BK7) comprises mostly silica (> 60%) with significant amounts of boron oxide, sodium oxide, and potassium oxide (> 10% each). There are also small amounts of other materials: barium oxide, zinc oxide, aluminum oxide, titanium oxide, and antimony oxide. The diversity of metal oxides promotes amorphous structure and neutral color. However, impurities still create fluorescence and scattering. D263 by SCHOTT is an extremely pure form of borosilicate for microscope slides and cover slips. Optical borosilicate crown is substantially free of iron oxide. N-BK7 is based on titanium and antimony, while BK7 is based on lead and arsenic.

An extremely pure form of borosilicate BSL7Y employs lead and arsenic oxide for removal of impurities. The lead reduces viscosity, while the arsenic removes iron. However, BSL7Y is 10 times more expensive than ecofriendly N-BK7. Other lead glass types, such as PBM27, are essential for exposure to UV and x-ray wavelengths. The purity of lead glass with arsenic greatly exceeds the purity of titanium glass with antimony. Furthermore, lead glass does not crystallize, while titanium glass does crystallize easily.

17.2 Glass Map

The glass map (Fig. 17.2) indicates refractive index versus Abbe number [Eq. (5.11)]. In Chapter 14, a simplified version of the glass map for the Erfle eyepiece was shown in Fig. 14.7. The glass curve describes the progression from traditional crown glass through a range of flint glass as the concentration of lead oxide increases. A material to the left of the glass curve is considered a fluorite glass because these glasses were originally based on fluorite. As the Abbe number increases, the dispersion decreases. Ergo, the Abbe number represents the reciprocal dispersion.

Figure 17.2 displays today's common types of SCHOTT glass. The labels are based on German words as indicated in parenthesis in the following text. A crown (or krone) has a low dispersion, whereas a flint (flint) has a high dispersion. A heavy flint (or schwer flint) contains more lead oxide than a light flint (or licht flint). The separation between crown and flint may be defined at an Abbe number of 50. The glass curve describes the progression from crown glass into flint glass as the concentration of lead oxide increases. Titanium oxide has largely replaced lead oxide for environmental benefits. Antimony has replaced arsenic as a fining agent, which reduces bubbles and impurities.

Lanthanum glass provides a high index with low dispersion. Its inner-shell electrons are tightly bound, which exhibit low dispersion. Its outer-shell electrons are tightly bound to oxygen. Thus, lanthanum's electron structure resembles that of silica, but there are many more electrons. Consequently, the dispersion of lanthanum glass is similar to silica while the refractive index is much larger. Lanthanum glass provides a high-index alternative to fluorite glass. The optical properties of numerous glass types are listed in Table 17.3.

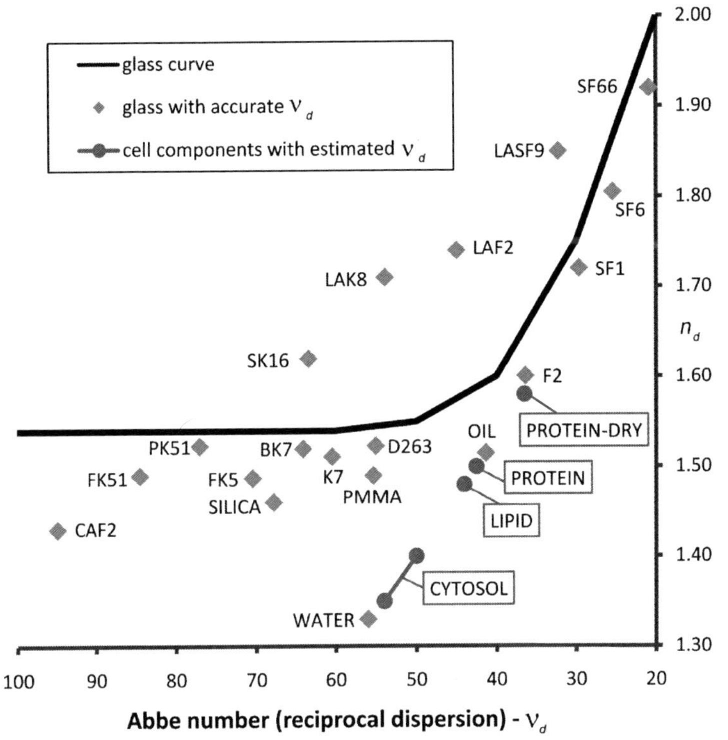

Figure 17.2 Glass map for common materials. A diamond indicates accurate values for both the refractive index and the Abbe number. A circle indicates an accurate refractive index with an estimated Abbe number. The linear plot of circles is based on water and TYPE A oil.

17.3 Fluorite

Fluorite glass can involve several materials: calcium fluoride (CaF_2), phosphate krone (PK), and fluorophosphate krone (FK). Phosphate krone and fluorophosphate krone qualify as glass because they are amorphous. However, calcium fluoride is not a glass; it is a single crystal. In the field of chemistry, fluorite indicates calcium fluoride (CaF_2); without exception, fluoride indicates the anion F^-, and fluorine indicates the element F. Figure 17.3 displays a periodic table with common elements of fluorite glass.

The tightly bound electrons of fluorite glass define an absorption band that is located deep in the UV regime. Consequently, a fluorite glass is effective for UV excitation during fluorescence. Figure 17.4 displays a glass map of suitable glass for application to fluorescence.

Calcium fluoride has a refractive index of 1.43 and an Abbe number of 95. It is located at the far left of the glass curve. It has the same number of 2p electrons as SiO_2. However, the electrons of fluorite are more tightly bound due to the increased charge of the fluorine nucleus. Consequently, the resonance is shifted to a higher frequency [Eq. (19.11)] and a shorter wavelength [Eq. (17.1)].

Table 17.3 Optical properties of common glass.

Type	Label	Glass name†	n_d	V_d	λ_{min}††	Λ_{max}††	Relative cost
Crown	BK7	Borosilicate krone	1.52	64	330	2000	1
Crown	PMMA	Polymethylmethacrylate	1.49	57	~365~	~1060~	0
Crown	SILICA	Fused silica, NIFS-S	1.46	68	~350~	~3500~	5
Crown	SILICA	Fused silica, NIFS-U	1.46	68	~250~	~3500~	10
Crown	SILICA	Fused silica, NIFS-A	1.46	68	~190~	~3500~	30
Flint	F2	Flint	1.62	36	390	2000	2
Flint	SF1	Schwer flint	1.72	30	410	2000	4
Flint	SF6	Schwer flint	1.81	25	420	2000	4
Flint	SF66	Schwer flint	1.92	21	460	2000	16
Fluor	FK51	Fluorophosphate krone	1.48	85	370	2000	26
Fluor	LaK8	Lanthanum krone	1.71	54	370	2000	3
Fluor	LaF2	Lanthanum flint	1.74	45	390	2000	4
Fluor	LaSF9	Lanthanum schwer flint	1.85	32	420	2000	7
Fluor	PK51	Phosphate krone	1.53	77	310	2500	27
Fluor	PSK53	Phosphate schwer krone	1.55	63	365	2000	7
Fluorite	CaF	Calcium fluoride - NICF-U	1.43	95	~225~	~6000~	50
Fluorite	CaF	Calcium fluoride - NICF-A	1.43	95	~175~	~6000~	100
Fluorite	CaF	Calcium fluoride - NICF-V	1.43	95	~125~	~6000~	400
Fluorite	FK5	Fluorophosphate krone	1.49	70	310	2000	2
Fluorite	FK51	Fluorophosphate krone	1.49	85	340	2500	18

† **German** – English: **krone** – crown; **schwer** – 1. heavy, 2. dense.
†† >90% at 10 mm thick.
~estimate.

The impurities of natural fluorite emit fluorescence; therein originates the word fluorescence. Synthetic fluorite (a single crystal of calcium fluoride) offers high levels of purity without fluorescence.

Calcium fluoride has a similar refractive index to borosilicate, but with a much higher Abbe number. A positive element of fluorite in combination with a negative element of short flint is an effective design for color correction in the blue wavelengths.

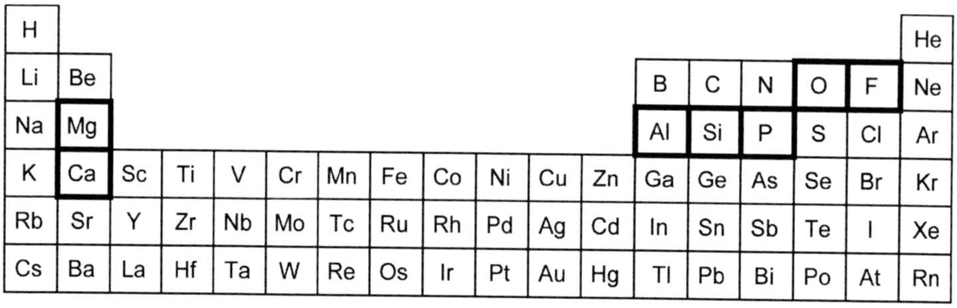

Figure 17.3 Periodic table of common elements in fluorite glass.

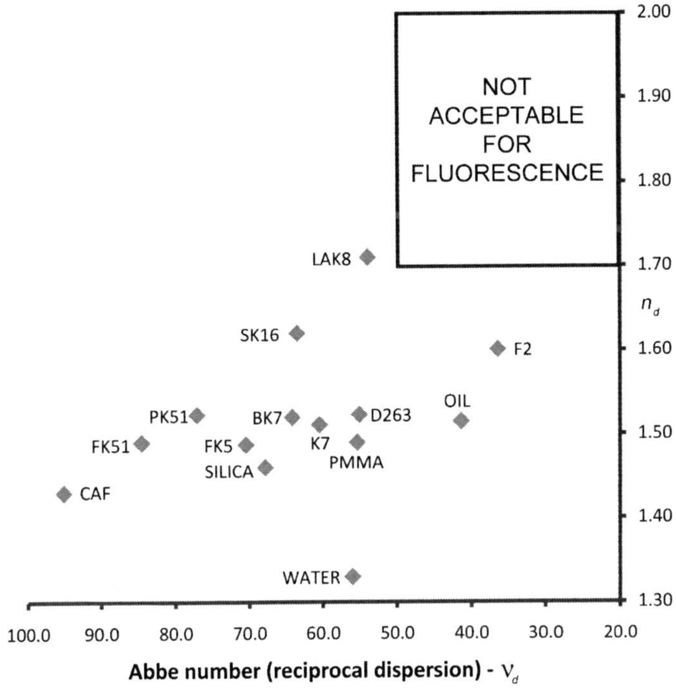

Figure 17.4 Glass map of materials acceptable for fluorescence. Suitable glass for fluorescent applications: $n_d < 1.70$ or $\nu_d > 50$. The intrinsic fluorescence and/or absorption of the upper right region is not acceptable in fluorescent applications. Acceptable region defined by Misawa.[22]

Calcium fluoride is an excellent material for UV applications. It may effectively correct chromatic aberration in combination with fused silica. However, the UV and deep-UV grades of these materials are extremely expensive. Calcium fluoride has a large CTE. The CTE of CaF_2 is 2.6 times that of BK7, and 33 times that of fused silica. It is prone to cracking from thermal shock.

Fluorophosphate glass employs phosphate and silica as the base network with the addition of fluorine. Phosphate krone (PK) is a mixture of phosphate and silicate. Phosphate binds electrons more tightly than silica. The addition of fluorine in place of oxygen creates fluorophosphate krone (FK). Fluorine binds electrons more tightly than oxygen. Consequently, the resonant frequency of FK extends further into the UV. The dispersion of BK or FK is less than that of silicate.

An FK displays a positive relative partial dispersion ΔP_{gF}. It is frequently employed for correction of the secondary color spectrum between the g and F lines. The positive relative partial dispersion of a fluorophosphate krone is combined with negative relative partial dispersion of a short flint.

Fluorophosphate glass can present difficult issues. The material can melt during polish. It reacts to atmosphere. A large CTE can promote cracking. The CTE of PK51 is 1.7 times that of BK7. The CTE of CaF_2 is 2.6 times that of BK7 and 33 times that of fused silica. Fluorophosphate glass is difficult to manage but well worth the effort.

17.4 Short Flint

A short flint displays a transmission spectrum with reduced transmission in the blue regime. The short spectrum length is created by a broadened absorption-band peak in the UV regime. The small absorption in the blue regime is created by resistance to dipole current. This resistance reduces the refractive index in the blue regime but not in the green or red regimes. The partial dispersion P_{gF} in the blue is smaller in a short flint than a normal glass. This defines a negative relative partial dispersion ΔP_{gF}.

There are several types of short flint. A kurz flint [(KzF), short flint] employs antimony in place of lead. A kurz flint sonder [(KzFS), short-flint special] employs a special glass such as borate (B_2O_3) in place of silica (SiO_2). The term "special" indicates a base network other than silica. A short-flint special KzFS of the 1900s is based on lead borate, whereas a normal flint is based on lead silicate. A lead-borate glass is reactive to the atmosphere.

Because of its importance as a short flint, the use of lead-borate glass persists today; however, there are new materials in development. A niobium silicate offers a short spectrum without the hazards of a lead-borate glass. However, the anomalous partial dispersion of niobium silicate is normally less than that of a lead borate.

Chapter 18 contains more details on short flint and short-flint special glass.

17.5 Anomalous Dispersion

Figure 17.5 displays the refractive indices of glass with both normal and anomalous dispersion. The relative partial dispersion ΔP_{gF} is an important metric of glass, as it indicates the relative optical power in the blue spectrum (g to F) to the optical power in the red spectrum (d to C). Section 19.3 provides a mathematical definition of ΔP_{gF}.

A normal glass is defined as the flint F2. The refractive index of a normal glass grows faster in the g-to-F spectrum than in the d-to-C spectrum. An anomalous dispersion departs from the slopes of F2.

The refractive index of dense flint SF2 grows more rapidly in the g-to-F spectrum. Thus SF2 displays a positive relative partial dispersion ΔP_{gF}.

The refractive index of a short flint KzFH1 grows at the normal rate in the d-to-F spectrum; however, it grows more slowly in the g-to-F spectrum. Thus, a short flint defines a negative relative partial dispersion in the g-to-F spectrum. A short flint effectively corrects secondary color in the blue spectrum.

The refractive index of a lanthanum crown LaK21 grows more slowly over all wavelengths. Thus, LaK21 has a large Abbe number. It also displays a negative relative partial dispersion. However, a lanthanum crown is less effective at color correction than a short flint because the slope of a lanthanum crown is flatter in both the blue and red spectra.

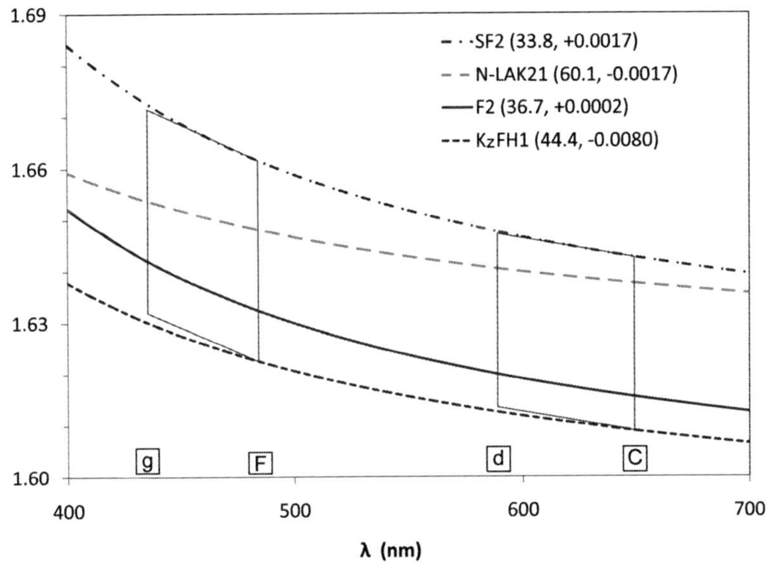

Figure 17.5 Refractive index of normal and anomalous glass. Glass types are indicated by name (Abbe number and relative partial dispersion). The flint F2 defines a normal relative partial dispersion near zero. The dense flint SF2 displays a positive relative partial dispersion. The short flint KzFH1 displays a negative relative partial dispersion. The lanthanum crown N-LaK21 displays a large Abbe number and a negative relative partial dispersion. The parallelograms indicate the slope of the F2 plot.

17.6 Sellmeier Formula

The refractive index of a normal glass is modeled as the summation of two electron states. A first resonator represents the bulk silicon dioxide. A second resonator represents the flint component, lead oxide or titanium oxide. A sum of the first and second resonators defines normal dispersion. The right portion of the glass curve also represents normal dispersion. Addition of a third resonator, such as lanthanum oxide, may create anomalous dispersion. Table 17.4 lists the Sellmeier parameters for the glass of Fig. 17.5.

The refractive index of three resonators is expressed by the Sellmeier formula as derived from the Lorentz Oscillator [Eqs. (19.9) and (19.11)]:

$$n^2 - 1 = \left(\frac{K_1 \lambda^2}{\lambda^2 - L_1} \right) + \left(\frac{K_2 \lambda^2}{\lambda^2 - L_2} \right) + \left(\frac{K_3 \lambda^2}{\lambda^2 - L_3} \right). \qquad (17.1).$$

The normal dispersion of F2 is defined by the first and second resonators. A large L_3 minimizes the effects of the third resonator in the visible regime. The resonant wavelengths for the first and second oscillators are similar, as indicated by L_1 and L_2. The concentrations of the first and second oscillators are indicated by K_1 and K_2. SF2 displays a larger concentration of both oscillators by 5 to 10% over F2. Consequently, SF2 displays a large index. The similar resonant wavelengths indicate similar dispersion in F2 and SF2.

Table 17.4 Sellmeier parameters of normal and anomalous glass.

Sellmeier	F2	SF2	N-LaK21	Units
K1	1.345	1.403	1.227	per μm^2
L1	0.010	0.011	0.006	μm^2
K2	0.209	0.232	0.421	per μm^2
L2	0.047	0.049	0.020	μm^2
K3	0.937	0.939	1.013	per μm^2
L3	112	112	88	μm^2

SCHOTT	KzFH1	Units
A0	$2.55 \times 10^{+00}$	—
A1	-1.29×10^{-02}	per μm^2
A2	1.84×10^{-02}	μm^2
A3	5.86×10^{-04}	μm^4
A4	-1.62×10^{-05}	μm^6
A5	2.27×10^{-06}	μm^8

The anomalous dispersion of LaK21 is defined by shorter resonant wavelengths. This indicates a resonance deeper into the UV, which is beneficial to UV excitation. Also, the second resonant wavelength is shorter than that of F2, while the concentration has doubled. The visible spectrum now extends farther from the UV.

The dispersion of glass normally indicates the location of absorption bands. A low-dispersion glass indicates a short resonant wavelength without absorption in the blue. A high-dispersion glass indicates a near-UV resonant wavelength with some absorption in the blue. A high-dispersion glass might appear yellow due to absorption in the blue.

The anomalous dispersion of the short flint KzFH1 requires the SCHOTT formula as follows:

$$n^2 = A_0 + A_1\lambda^2 + \frac{A_2}{\lambda^2} + \frac{A_3}{\lambda^4} + \frac{A_4}{\lambda^6} + \frac{A_5}{\lambda^8}, \tag{17.2}$$

which is a mathematical formula without physical significance. However, it can be essential for accurate description of anomalous dispersion. Table 17.4 displays the SCHOTT parameters for KzFH1.

Anomalous partial dispersion is essential for correction of chromatic aberration. Table 17.5 displays the Sellmeier coefficients of silica and fluorite glass. The resonant wavelengths are smaller, and the silica and fluorite glass display far less dispersion than a normal glass. However, they are more expensive in cost.

Table 17.5 Sellmeier parameters of silica and fluorite glass.

Sellmeier	Silica	PK51	FK51	CaF_2	Units
K1	0.696	1.156	0.971	0.568	per μm^2
L1	0.005	0.006	0.005	0.003	μm^2
K2	0.408	0.153	0.217	0.471	per μm^2
L2	0.014	0.019	0.015	0.010	μm^2
K3	0.897	0.786	0.905	3.848	per μm^2
L3	98	141	169	1,201	μm^2

The relative partial dispersion for g and F is frequently cited as ΔP_{gF}. A fluorite krone displays a large and positive ΔP_{gF}. A schwer flint displays a small positive ΔP_{gF}. A short flint KzF displays a small negative ΔP_{gF}. A short-flint special displays a large negative ΔP_{gF}. An lanthanum krone or lanthanum flint displays a negative ΔP_{gF}. The positive ΔP_{gF} of a fluorite krone is frequently combined with the negative ΔP_{gF} of a short flint.

17.7 Environmentally Safe Glass

In 2002, the European Community issued two environmental directives on the use of lead and other substances. Lead is an important component of high-index flint and short-flint special glass. Lead is even contained in some forms of brass.

The directive on Waste Electrical and Electronic Equipment 2002/96/EC defines ten categories of electrical and electronic equipment. Two categories are quite relevant to microscopes: medical devices and lighting equipment. Standards for reuse, recycle, and disposal are defined.

The directive on Restriction of Hazardous Substances, 2002/95/EC, restricts the use of six materials in electronic components and assemblies. Specifically, these materials are lead, mercury, cadmium, hexavalent chromium, polybrominated biphenyls, and polybrominated diphenyl ethers. Exceptions are permitted when lead is necessary.

In most flint glass, lead oxide has been replaced by titanium oxide or niobium oxide. The absorption of titanium oxide extends farther into the blue than lead oxide. Titanium crystallizes more easily than lead oxide. The crystalline inclusions of an ecoglass can be pulled from the glass during manufacture. The "grind and shine" of ecoglass can be difficult, as the pulled inclusions create scratches.

Some short flint glasses require lead borate for their unique partial dispersion in the blue. A high-index glass with low anomalous partial dispersion is essential for many instruments. A niobium silicate may replace some lead borate but not all.

Numerous challenges exist in the conversion to ecoglass. The demand for cell phone cameras is pushing the further development of glass. Jademzik et al. (of SCHOTT glass) describe the challenges and benefits of ecofriendly glass.[50]

17.8 Glass Code

A glass code defines a glass by its refractive index at the d line and its Abbe number. The first three digits apply to the first three decimals of the refractive index. The second three digits apply the first three digits of the Abbe number. Thus, the glass code for K7 (1.511, 60.4) is 511-604. Other common glass codes are 517-642 for BK7 (1.517, 64.2) and 620-363 for F2 (1.620, 36.3).

Manufacturers strive to provide identical glass types by glass code; however, there are significant differences. The anomalous partial dispersions might not be the same. Resistance to the atmosphere might vary. Other variations might apply.

17.9 Spectral Lines

An i-line glass indicates a tolerance to a UV wavelength of 365 nm. There are numerous formats of i-line glass that are ecofriendly. They are based on titanium oxide and antimony. An i-line glass is much more expensive than ecofriendly borosilicate.

Ordinary borosilicate transmits at 330 nm (Table 17.3); however, there is a significant absorption of 3% per 25 mm of glass at the i line (365 nm). A small absorption of illumination can create substantial background fluorescence and/or browning.[i]

Lead glass might be required for x-ray exposure (31 keV, 40 pm). Lead glass has two distinct advantages in purity over ecoglass: reduced viscosity and lack of titanium. The reduced viscosity of lead glass facilitates removal of impurities. The absence of titanium prevents tight binding of impurities within titanium-oxide crystals. Furthermore, arsenic promotes removal of impurities.

Other wavelengths of interest are listed in Table 17.6. The e line is very close to the peak sensitivity of the human eye at 555 nm. The F′ and C′ lines shift the F and C lines toward a better match to human vision. An optical design for human vision is frequently based on the F′, d, and C′ lines.

17.10 Cost of Optics

There are two important components of cost in fabrication of optics: quality of material and ease of processing. The cost of the bulk material is largely driven by a few factors: bubbles, inclusions, striae, and purity. The cost of processing is driven by numerous factors: surface-form errors, surface defects, cleaning, coating, etc. Table 17.3 displays various glass parameters and relative cost of BK7.

---

[i] "Browning" indicates discoloration due to transformation of the glass.

Table 17.6 Spectral lines of emission.

Line	Wavelength (nm)	Source
i	365.0	Hg
h	404.7	Hg
g	435.8	Hg
He-Cd	441.6	Laser
F′	480.0	Cd
F	486.1	H
Ar	514.5	Laser
e	546.1	Hg
d	587.6	He
He-Ne	632.8	Laser
C′	643.9	Cd
C	656.3	H

Ecofriendly glass can display issues with inclusions and impurities. Lead and arsenic are more effective than titanium and antimony for elimination of impurities. Consequently, an ecofriendly glass might not tolerate UV or x-ray exposure, while its lead-oxide predecessor tolerates exposure without issue. Titanium oxide may crystallize in the form of small inclusions that leave scratches during polish. Impurities may also create intrinsic fluorescence.

17.11 Structural Materials

Table 17.7 lists the quality and cost of common materials within an optical instrument. The CTEs of doublets must be considered for large temperature ranges. The higher CTE glass may crack during rapid change in temperature. The density of glass is an important tool for experimental determination of a glass type.

Stainless steel contains mostly iron with essential amounts of nickel and chromium. Nickel promotes a face-centered cubic structure, which is more flexible than other structures. Chromium creates a passive-oxide layer, which does not grow into the material. Types 304 and 316 stainless steel are recommended for biological applications. The natural chrome-oxide layer is an acceptable finish. A black iron-oxide treatment may reduce reflectance. A small amount of carbon (0.1%) strengthens stainless steel by bridging gaps in a similar manner to boron in glass.

A lens barrel and retainer ring should be made of different metals. Stainless steel provides a strong material for the threads of the barrel. Brass provides a softer material for the threads of the retainer. The retainer might even deform by design on contact with the lens. Brass will not stick to stainless steel, whereas similar types of steel will fuse under friction and pressure.

Table 17.7 Structural properties of common materials.

Label	Name (composition†)	Density (g/cm^3)	CTE (10-6/°C)	Tensile†† (MPa)	Cost in 2008 ($/kg)	Cost in 2008 ($/L)
BK7	Borosilicate krone	2.5	7.1		3	8
PMMA	Polymethylmethacrylate	1.2	80.0	50	1	1
NIFS-S	Fused silica up to 350nm	2.2	0.6	48	21	46
F2	Flint	2.7	7.8	—	14	37
SF1	Schwer flint	3.0	9.1	—	28	85
SF6	Schwer flint	3.4	9.0	—	25	83
SF66	Schwer flint	4.0	5.9	—	112	448
LaK8	Lanthanum krone	3.6	5.6	—	21	75
LaF2	Lanthanum flint	4.3	8.1	—	25	105
LaSF9	Lanthanum schwer flint	4.4	7.4	—	49	216
FK51	Fluorophosphate krone	3.7	13.3	—	182	679
PK51	Phosphate krone	3.6	12.4	—	189	673
NICF-V	Calcium fluorite up to 225 nm	3.2	18.4	—	70	223
FK5	Fluorite krone	2.5	9.2	—	14	36
FK51A	Fluorite krone	3.7	12.7	—	126	464
304	Stainless steel 18-8 (Fe, Cr, Ni)	8.0	17.2	586	13	107
316	Surgical stainless steel (Fe, Cr, Ni)	8.0	15.9	621	20	163
416	Ferric stainless steel (Fe, Cr)	7.8	9.9	655	32	248
4340	Chromoly steel (Fe, Cr, Mo)	7.8	11.5	1,241	21	163
1095	Blue spring steel (Fe, C)	7.8	11.4	979	97	763
2024	Aircraft aluminum (Al, Cu, Mg)	2.8	22.8	441	111	306
6061	Aluminum (Al, Mg)	2.7	23.4	241	53	143
360	Brass (Cu, Zn, Pb)	8.5	11.4	400	19	158
—	Northern red oak, dry	0.6	—	98	2	0.1

† Components at > 1% by weight
†† 1 MPa = 10 kg/cm^2

Aluminum 6061 is a common alloy for optical fixtures. It is softer than stainless steel and easier to machine. However, it must be anodized, otherwise an aluminum-oxide dust will form and distribute on the optics. Aircraft aluminum 2024 contains copper for additional strength, which is not likely necessary for an optical instrument. MIC-6 is a aluminum alloy with minimal internal stress that permits precise machining.

Chapter 18
Composition and Spectra of Materials

18.1 Glass Structure

The descriptions of glass types in this book are largely based on the descriptions by Clement.[51] The Sellmeier coefficients are derived from vendor data as cited in ZEMAX™. The wavelength increment of the Sellmeier index is 1%. The transmission data are also derived from vendor data over 25 mm of internal transmission, as cited in ZEMAX.

A glass is a random network of fused inorganic material. It may be considered the amorphous form of a ceramic. Crystallization is defeated by a sufficient rate of cooling for the viscosity of the glass. A glass with thick viscosity may cool slowly due to the poor mobility of the components. A glass with low viscosity must cool rapidly before crystals grow.

18.2 Crown

A krone, or crown (K), is basically soda lime glass:

$$SiO_2\text{—}CaO\text{—}Na_2O, \tag{18.1}$$

where silica (SiO_2) is derived from sand, sodium oxide (Na_2O) is derived from soda, and calcium oxide (CaO) is derived from lime. Soda lime glass might also contain potassium oxide (K_2O), which is derived from potash (K_2CO_3). The silicate is a glass network former. Sodium oxide and calcium oxide are network modifiers that lower the melting point and the viscosity, respectively. Diversity in glass defeats crystallization and promotes homogeneity in the glass melt. Silica is the base network of most glass.

A borosilicate krone (BK) is frequently described as

$$SiO_2\text{—}B_2O_3\text{—}M_2O, \tag{18.2}$$

where the alkali metal oxide (M_2O) is a network modifier, such as sodium oxide (Na_2O) and potassium oxide (K_2O). The addition of the alkali oxides lowers the melting point and reduces the viscosity; these features promote ease of glass manufacture. The addition of borate (B_2O_3) fills voids and reduces the thermal coefficient of expansion. Borate is a network modifier in smaller concentrations, as shown in Table 18.1.

18.3 Flint

A flint (F) contains lead oxide (PbO) and alkaline oxide (M_2O):

$$SiO_2\text{—}PbO\text{—}M_2O. \tag{18.3}$$

Flint[i] occurred naturally as flint stone from the south coast of Britain. Lead oxide provides higher electron density than silica. Lead oxide also shifts UV resonance toward the visible regime. A high lead content can make glass appear yellow. Lead oxide is weakly reactive to the atmosphere. For improved chemical stability, titanium oxide (TiO_2) and zirconium oxide (ZrO_2) may replace small amounts of lead oxide at the cost of increased absorption in the blue. Figure 18.1 displays the Sellmeier index of a crown glass (K7) and flint (F2). The addition of lead oxide shifts UV resonance toward the blue. The F2 glass displays a larger partial dispersion ΔP_{gF} than the K7. The shift of resonance toward the blue also shortens the transmission spectrum (Fig. 18.2).

Lead oxide has several important features as a network modifier. The 90 electrons of lead oxide provide a higher electron density than the 30 electrons of silica (SiO_2). This increases the refractive index. The large mass of lead (82 u) has a lower thermal velocity than the silica (14 u) within a glass melt. Consequently, the lead oxide defeats crystallization through its low mobility. The

Table 18.1 Composition of borosilicate glass S-BSL7 (Ohara MSDS 2009).

Compound	Formula	Minimum	Maximum
Silica	SiO_2	65	75
Potassium oxide	K_2O	10	20
Sodium oxide	Na_2O	10	20
Boron oxide	B_2O_3	5	15
Barium oxide	BaO	—	5
Zinc oxide	ZnO	—	5
Aluminum oxide	Al_2O_3	—	1
Titanium oxide	TiO_2	—	0.5
Antimony oxide	Sb_2O_3	—	0.3

[i] Flint is derived from *splie*, meaning split. Gemanic *fli-, or spli-*.

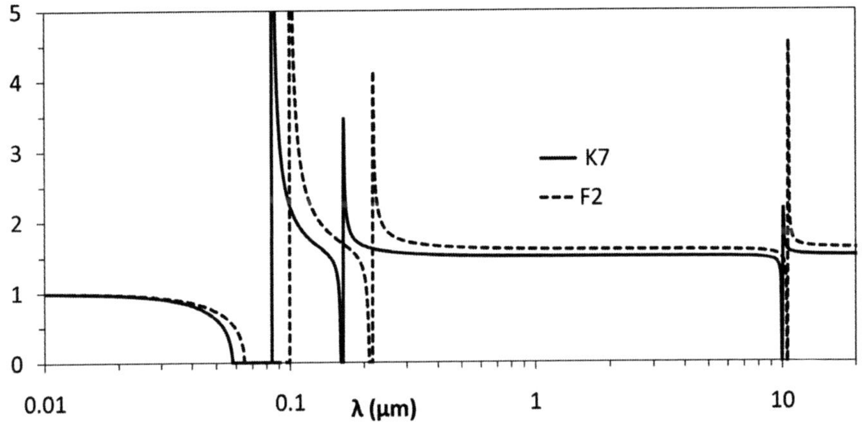

Figure 18.1 Sellmeier spectra of K7 and F2. SCHOTT defines normal partial dispersion by these two glass types. The addition of lead shifts the resonant peak toward the visible.

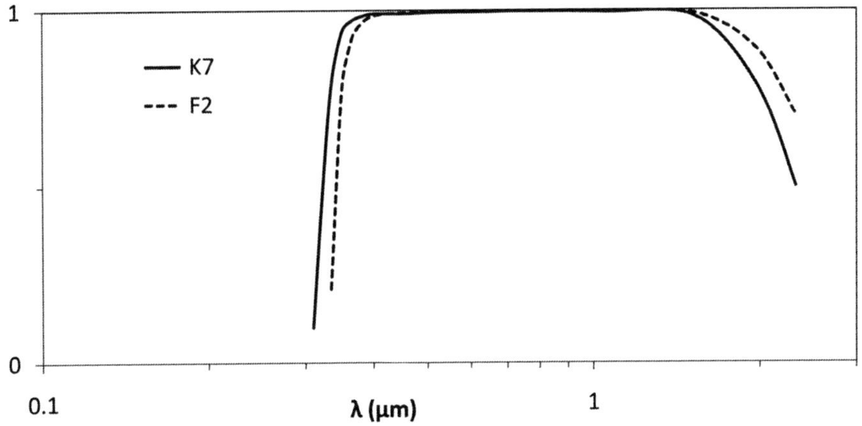

Figure 18.2 Transmission spectra of K7 and F2. The addition of lead shortens the spectrum in the blue.

bond length of lead oxide at 4.0 Å is much longer than silica at 1.6 Å. Consequently, its weaker bond promotes lower viscosity. To the benefit of the art of glass blowing, lead oxide has two important effects: its high index creates sparkle, and its high mass extends the working temperature.

Lead is the most common heavy metal in the earth's crust. Three of four heavy atoms spontaneously decay into a stable form of lead. Its high electron density creates a bright reflection in a thin coating of paint. White lead ($PbCO_3$) is the basic pigment of lead paint, which is now strictly regulated and controlled in numerous countries. As existing lead paint deteriorates, it settles in soil and ground water. The same destination applies to the lead silicate in glass parts. Fine particles of lead and other heavy metals must be properly disposed. A few early symptoms of lead poisoning are diffuse muscle weakness, joint pain, nausea, diarrhea, and constipation. The effects of lead on the central nervous system are not, as yet, reversible.

18.4 Long Crown

A lang krone, or long crown (LgK), displays a long optical spectrum with a small Abbe number. In fluorite krone (FK), some oxygen is replaced by fluorine. A fluorophosphate glass composition is frequently described as

$$P_2O_5\text{—}Al_2O_3\text{—}MO\text{—}(F_2 - O), \tag{18.4}$$

where (F_2 – O) represents a "fluorine minus oxygen." The phosphate P_2O_5 is a network former. The alumina (Al_2O_3) is a network modifier that promotes stability, as the phosphate tends to react with atmosphere. The alkaline metal oxide (MO) acts as a network modifier. The fluorite component (F_2 – O) binds electrons more tightly, which shifts the resonance deeper into the UV. This shift increases the spectrum length and reduces the dispersion.

Figure 18.3 displays the Sellmeier index of a krone (K7) and a fluorite krone (N-FK5). The addition of fluorite shifts the UV resonance away from the blue. The fluorophosphate krone glass displays a smaller partial dispersion ΔP_{gF} than the K7. The fluorophosphate krone also displays a longer transmission spectrum (Fig. 18.4).

A fluorite glass is reactive to atmosphere. A coating such as magnesium fluoride (MgF_2) is required for protection from the atmosphere. It can also melt during grinding or polish due to its low glass temperature. It might also fracture easily. The benefits of a low-dispersion glass have a significant cost in production.

18.5 Short Flint

A kurz flint, or short flint (KzF), provides a short optical spectrum through replacement of lead oxide with antimony oxide (Sb_2O_3):

$$SiO_2\text{—}Sb_2O_3\text{—}B_2O_3\,. \tag{18.5}$$

The borate is a network modifier that shifts the resonant wavelengths to shorter wavelengths. The antimony oxide provides electron density.

A short flint (KzFH1) employs shorter resonant wavelengths than flints such as F2 in Fig. 18.5. The transmission spectrum of a short flint is shorter in Fig. 18.6. The shorter resonant wavelengths of the antimony oxide reduce partial dispersion while maintaining the reactive index. The shortened transmission spectrum is created by a shorter lifetime of resonance. As indicated by the Lorentz model of Figs. 5.3 and 5.4, a shorter lifetime of resonance has two effects: a broader absorption band and a flatter index profile. The broader absorption band shortens the spectrum range in the blue. The flatter index profile creates anomalous partial dispersion in the blue. The relative partial dispersion ΔP_{gF} of the KzFH1 is –0.008.

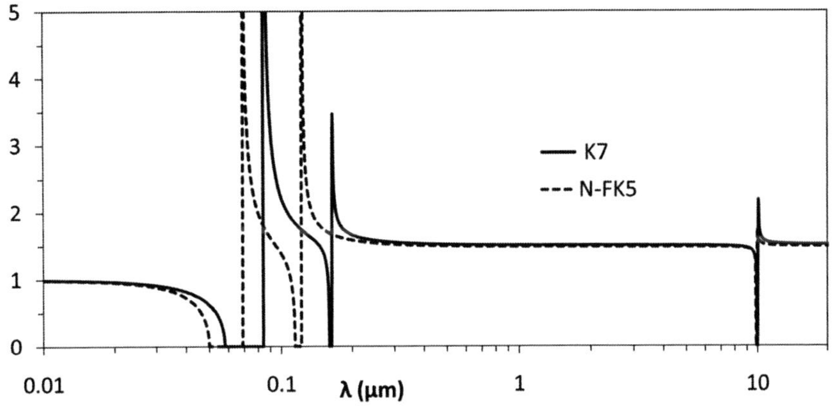

Figure 18.3 Sellmeier spectra of K7 and N-FK5. the addition of fluorite shifts the resonant peak from the visible regime.

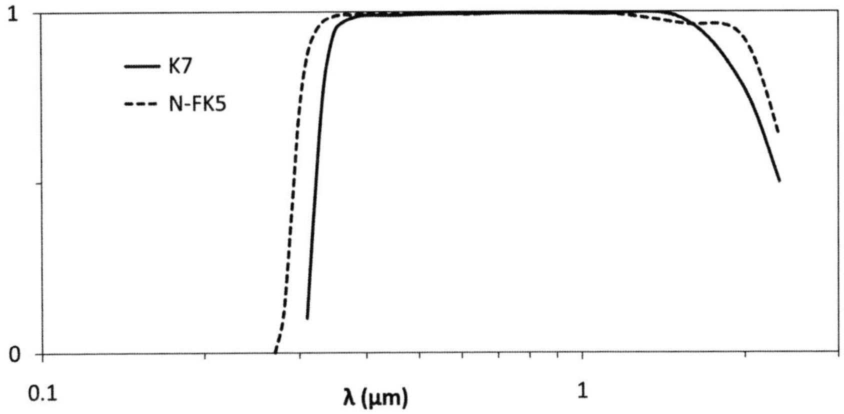

Figure 18.4 Transmission spectra of K7 and N-FK5. The addition of fluorite lengthens the spectrum in the blue.

18.6 Short-Flint Special

A kurz flint sonder, or short-flint special (KzFS), provides a short optical spectrum due to replacement of silica with borate as the network former:

$$B_2O_3\text{—}PbO\text{—}Al_2O_3\,. \tag{18.6}$$

The glass is frequently called lead borate. A special glass borate (B_2O_3) serves as the base network in place of silica. The alumina is a network modifier for chemical stability, while the lead silicate provides electron density. A short-flint special (KzFS) is reactive with the atmosphere. A protective coating is required. The lead borate of a short-flint special is reactive to the atmosphere. Storage in dry nitrogen is recommended until a protective coating is applied. Significant expertise is required for processing of a short-flint special glass. Ecofriendly versions of short-flint special are based on niobium and silicate.

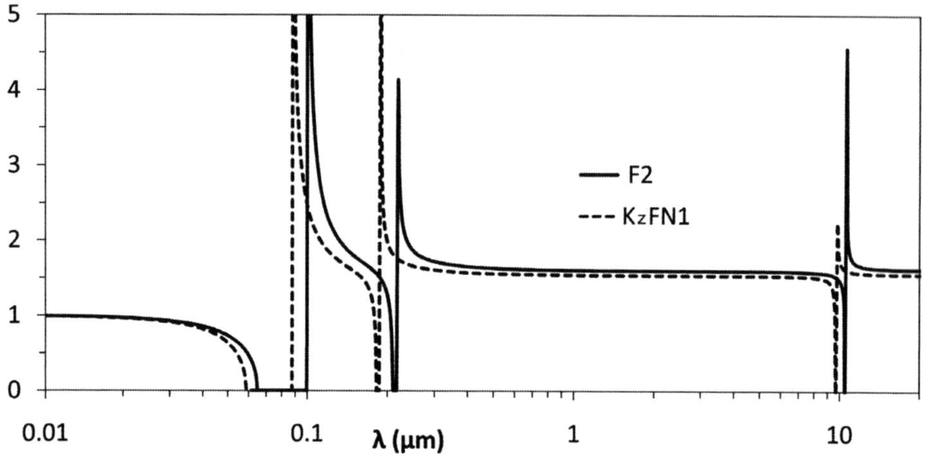

Figure 18.5 Sellmeier spectra of F2 and KzFS1. Antimony in place of lead shifts the resonant peak from the visible regime. A shortened lifetime flattens the refractive plot in the blue.

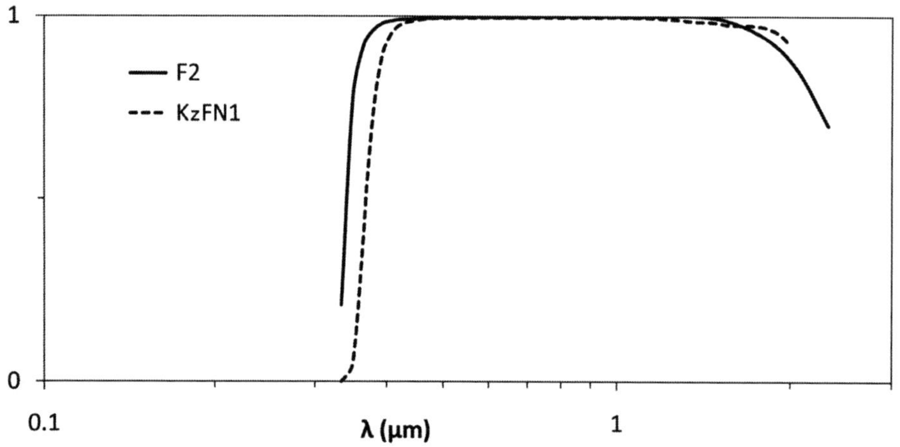

Figure 18.6 Transmission spectra of F2 and KzFS1. Antimony in place of lead shortens the spectrum in the blue.

The borate of the short-flint special KzFSN4 displays a resonance near 8 μm in Fig. 18.7. The absorption band of the borate KzFSN4 at 8 μm extends to 1 μm in Fig. 18.8, while the absorption band at 350 nm is not greatly widened. The shortened spectrum in the blue is enough to reduce the partial dispersion of the g and F lines. The relative partial dispersion ΔP_{gF} of the KzFSN4 is –0.009.

The partial dispersion of a short-flint special is reduced by a shorter lifetime. The broader absorption band shortens the spectrum range of the infrared from 10 to 8 μm. The flatter index profile creates anomalous partial dispersion in the blue.

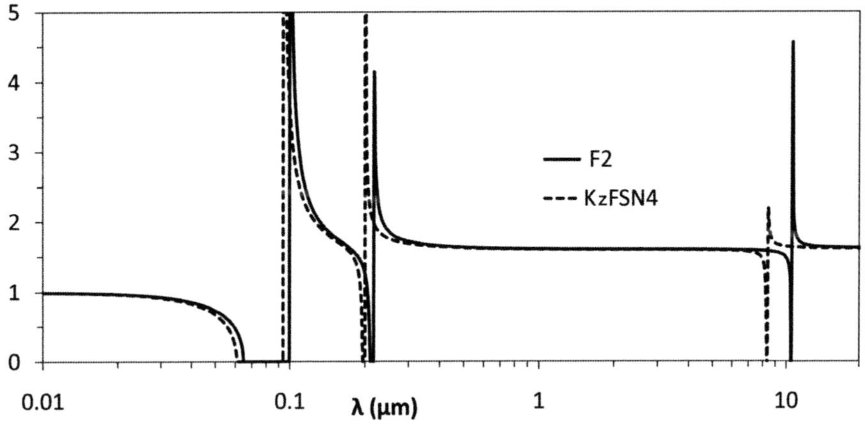

Figure 18.7 Sellmeier spectra of F2 and KzFS1. Lead borate in place of lead silicate shifts the resonant peak from the visible regime. A shortened lifetime flattens the refractive plot in the blue.

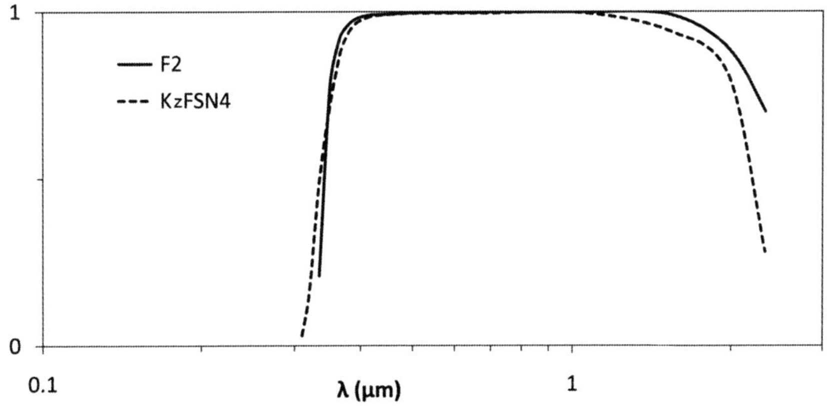

Figure 18.8 Transmission spectra of F2 and KzFS1. Lead borate in place of lead silicate shortens the spectrum weakly in the blue and strongly in the infrared.

The negative partial dispersion of a special short flint is essential for color correction in the blue wavelength range. A shorter lifetime of resonance decreases the partial dispersion. The relative partial dispersion is similar in magnitude to the short flint KzFSH1, while the transmission in the blue is much improved. A short-flint special provides a negative partial dispersion without a deep reduction of spectrum length in the blue.

18.7 Environmentally Safe Short Flint

SCHOTT still manufactures a lead borate short-flint special KzFSN5, which has ecofriendly versions: N-KzFS5 by SCHOTT and S-NBH5 by Ohara. Table 18.2 indicates the composition of S-NBH5, which is mostly silica and niobium oxide. The niobium oxide Nb_2O_5 is a network former of the same kind as the

Table 18.2 Composition of niobium silicate glass S-NBH5 (Ohara MSDS 2009).

Compound	Formula	Minimum	Maximum
Silica	SiO_2	31	41
Niobium oxide	Nb_2O_5	16	26
Boron oxide	B_2O_3	8	18
Sodium oxide	Na_2O	4	12
Zirconium oxide	ZrO_2	4	12
Zinc oxide	ZnO	3	9
Barium oxide	BaO	—	3
Calcium oxide	CaO	—	3
Potassium oxide	K_2O	—	3
Lithium oxide	Li_2O	—	3
Antimony oxide	Sb_2O_3	—	0.5

phosphate oxide P_2O_5. The zirconium oxide (ZrO_2) is a network modifier that shortens resonant lifetime of the niobium oxide. Zirconium has one less electron than niobium. The addition of zirconium in place of niobium creates a hole[ii] in the valance band of the niobium-oxide network. A hole provides opportunity for electron scatter and shorter lifetime of oscillation. The other metal oxides provide diversity toward a lower melting point, lower viscosity, and less crystallization. The antimony oxide removes iron.

The name of the Ohara glass S-NBH implies several features: S indicates an environmentally safe glass; NB indicates the bulk composition as niobium; H indicates a high index for this glass type.

A niobium-silicate short flint is very similar to a lead-borate short flint; however, it is also different. The refractive index of S-NBH5 nearly equals that of KzFSN5 in the visible regime, as shown in Fig. 18.9. The refractive indices at the d lines are the same at 1.65412. The Abbe numbers are similar at 39.68 for the S-NBH5 and 39.63 for the KzFSN5. The transmission of both materials is nearly identical from 436 to 550 nm. This indicates a similar reduction in resonant lifetime, which is essential for a short flint. However, there is significant difference in the relative partial dispersion: ΔP_{gF} equals –0.004 for S-NBH5, and –0.007 for KzFSN5. The smaller ΔP_{gF} of S-NBH5 also correlates with improved spectrum length in Fig 18.10. S-NBH5 displays a longer spectrum with less absorption in the blue.

In summary, the lead borate displays more relative partial dispersion, while the niobium silicate is much more chemically stable. Environmental stability

[ii]A hole is an empty electron state.

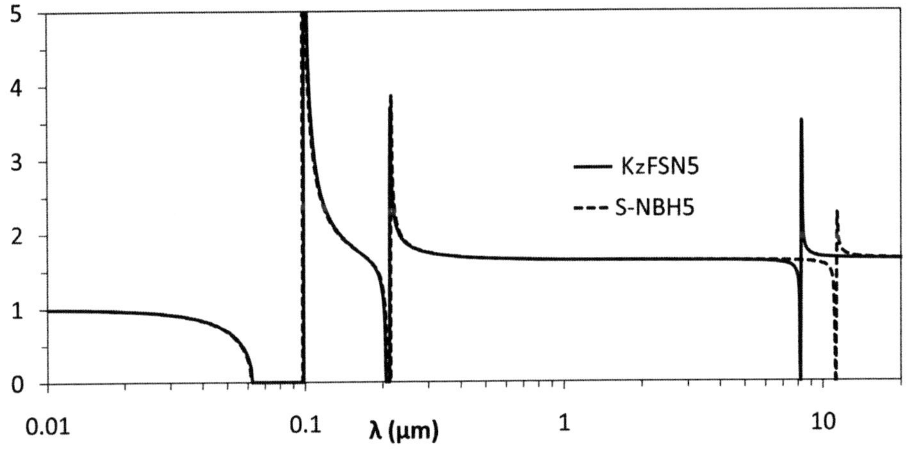

Figure 18.9 Sellmeier spectra of KzFSN5 and S-NBH5. Niobium silicate in place of lead borate replicates the spectrum in the visible regime.

normally requires some compromise in glass features. The holes responsible for electron scatter also promote chemical reaction. Absorption and reaction are intrinsic features of a short-spectrum flint with anomalous partial dispersion.

18.8 Dense Flint

A schwer flint, or dense flint (SF), employs a larger concentration of lead oxide than normal flint. The refractive index of SF6 is larger than that of F2 in Fig. 18.11. The addition of more lead oxide shifts the UV resonance further into the blue in Fig. 18.12.

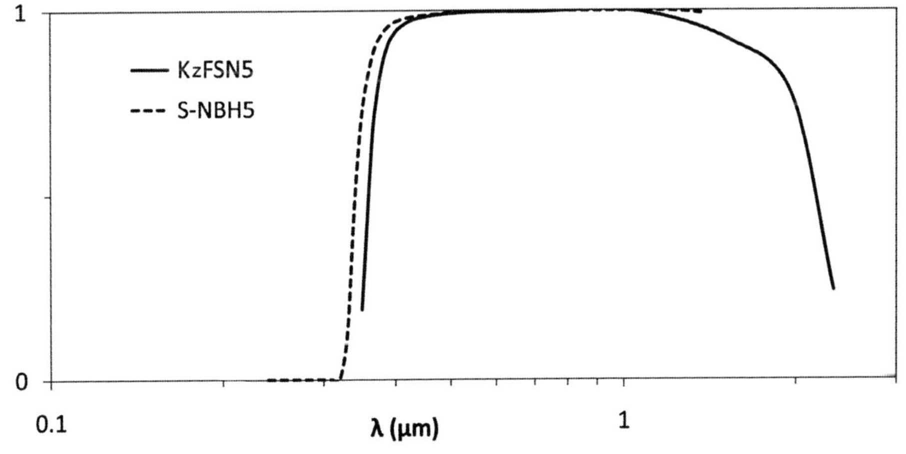

Figure 18.10 Transmission spectra of KzFSN5 and S-NBH5. Niobium silicate in place of lead borate extends the spectrum from the visible regime. Less absorption indicates longer lifetime and less negative relative partial dispersion.

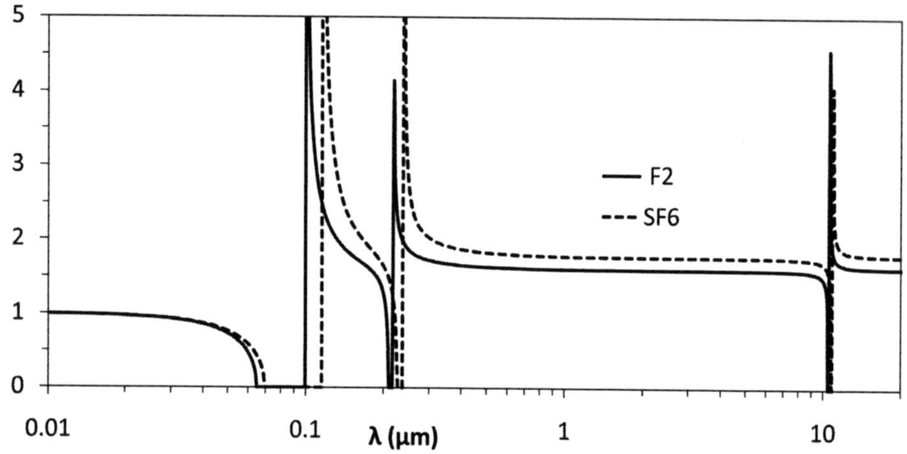

Figure 18.11 Sellmeier spectra of F2 and SF6. The addition of lead oxide in place of silica greatly increases refractive index and dispersion.

An ecofriendly version of schwer flint employs titanium oxide in place of lead oxide. There is a small shift of resonant wavelength into the blue, as shown in Fig. 18.13. The refractive-index profile is substantially the same. However the absorption peak of the titanium oxide flint extends far into the visible spectrum, as shown in Fig. 18.14. There is significant absorption from 400 to 580 nm. The ecofriendly N-SF6 can appear yellow. The composition of the ecofriendly version of SF6 is displayed Table 18.3. The S-TIH6 glass by Ohara indicates silica and titania oxide as the major components (25–35%). Zirconium oxide and niobium oxide defeat crystallization of the titanium oxide through diversity of similar bonds. Other components are the typical ingredients of borosilicate glass.

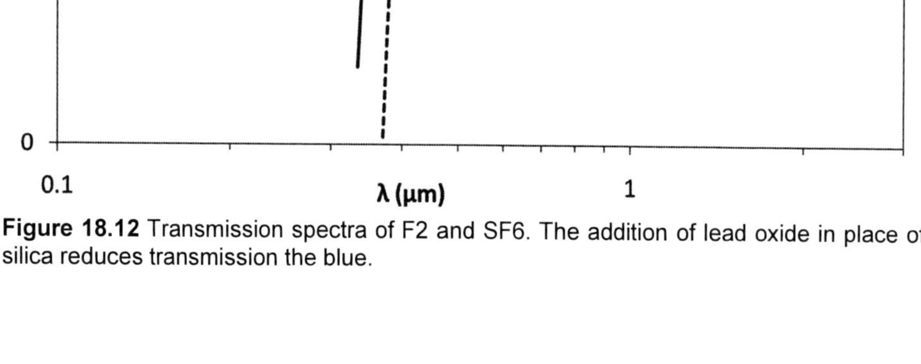

Figure 18.12 Transmission spectra of F2 and SF6. The addition of lead oxide in place of silica reduces transmission the blue.

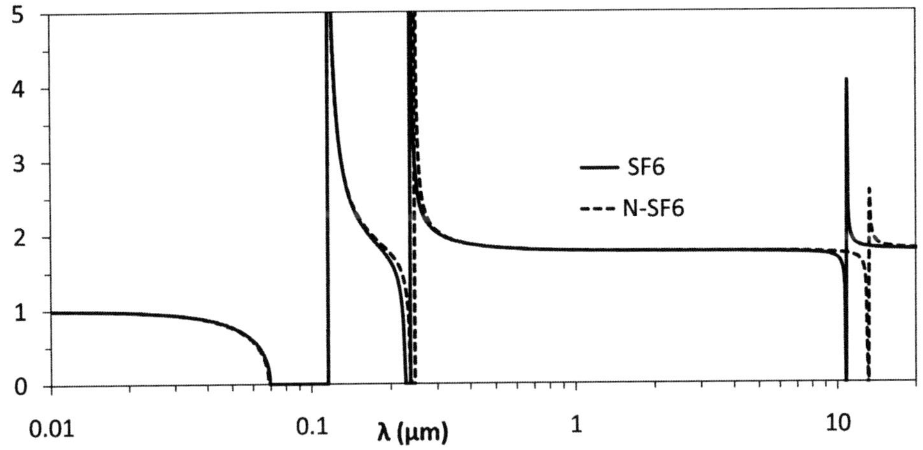

Figure 18.13 Sellmeier spectra of SF6 and N-SF6. Titanium oxide in place of lead oxide replicates the refractive index in the visible.

An important issue with ecofriendly glass is crystallization. Titanium oxide crystallizes more easily than lead oxide. The crystallites occur in numerous shapes: filaments, cords, and other inclusions. Filaments and cords create an inhomogeneous refractive index. Inclusions may scratch the surface as they are pulled during polish. Most lead glass can be replaced with titanium oxide; however, titanium glass is more difficult and more expensive to process.

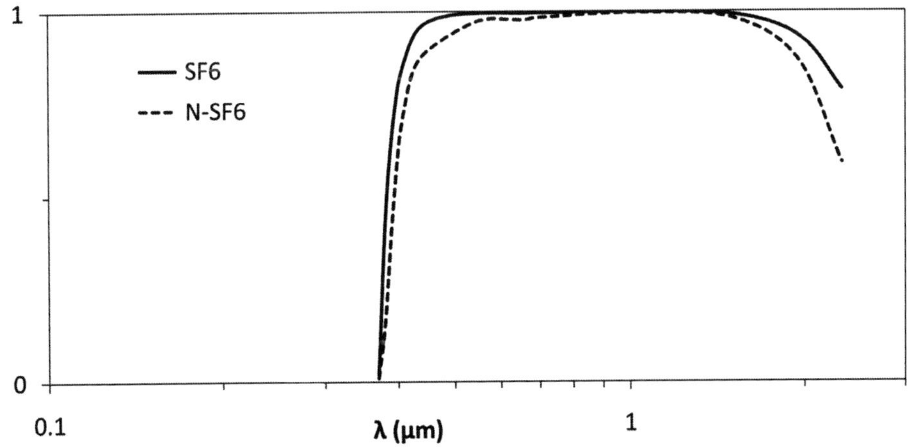

Figure 18.14 Transmission spectra of SF6 and N-SF6. Titanium oxide in place of lead oxide reduces transmission in the blue and green.

Table 18.3 Composition of titanium silicate glass S-TIH6 (Ohara 2009).

Compound	Formula	Minimum	Maximum
Silica	SiO_2	25	35
Titanium oxide	TiO_2	25	35
Barium oxide	BaO	10	20
Calcium oxide	CaO	10	20
Potassium oxide	K_2O	10	20
Sodium oxide	Na_2O	10	20
Niobium oxide	Nb_2O_5	5	15
Zirconium oxide	ZrO_2	—	5
Antimony oxide	Sb_2O_3	—	0.3

Chapter 19
Advanced Concepts

19.1 Wave Equation

An electromagnetic wave is governed by three important features: the spatial curvature of the electric field $\nabla^2\mathbf{E}$, the temporal curvature of the electric field $\ddot{\mathbf{E}}$, and the temporal slope of electric current density $\dot{\mathbf{J}}$, which may also be considered the acceleration of electric current density. These parameters are expressed in the wave equation as follows:

$$\nabla^2\mathbf{E} = \mu\left(\varepsilon_0\ddot{\mathbf{E}} + \dot{\mathbf{J}}\right), \tag{19.1}$$

where μ is the magnetic permeability and ε_0 is the electric permittivity.

The ∇ operator defines the gradient as

$$\nabla = \frac{\partial}{\partial x}\mathbf{i} + \frac{\partial}{\partial y}\mathbf{j} + \frac{\partial}{\partial z}\mathbf{k}, \tag{19.2}$$

where **i**, **j**, and **k** are unit vectors. The ∇^2 operator defines the Laplacian as

$$\nabla^2 = \frac{\partial^2}{\partial x^2}\mathbf{i} + \frac{\partial^2}{\partial y^2}\mathbf{j} + \frac{\partial^2}{\partial z^2}\mathbf{k}. \tag{19.3}$$

The temporal versions of the spatial operators are

$$\dot{\mathbf{A}} = \frac{\partial}{\partial t}\mathbf{A} \tag{19.4}$$

and

$$\ddot{\mathbf{A}} = \frac{\partial^2}{\partial t^2}\mathbf{A}. \tag{19.5}$$

Application of the spatial and temporal derivatives yields

$$k_x^2 + k_y^2 + k_z^2 = N^2 \frac{\omega^2}{c^2}, \tag{19.6}$$

where k_x, k_y, and k_z are the spatial frequencies along orthogonal axes, N is the complex refractive index, ω is the temporal frequency, and c is the speed of light. Each k^2 term represents a spatial curvature of the wave along an axis. The ω^2 term represents the temporal curvature of the wave.

Examination of Eq. (19.6) reveals an important concept: the sum of the spatial curvatures is proportional to the temporal curvature. The scale of spatial-to-temporal curvature is dependent on the refractive index. Thus, the spatial curvature is dependent upon the electric current.

19.2 Refractive Index

Free and bound electrons have very different effects on spatial curvature. The free electrons of a metal react quickly enough to cancel an external electric field. This creates an exponential decay of the electric field within a spatial wavelength (Fig. 19.1). The restoring force of bound electrons prevents complete cancellation of the external electric field. Consequently, the bound electrons drive the electric field faster toward zero but not completely to zero (see Fig. 19.2). The faster return to zero is described by larger spatial frequency or shorter wavelength.

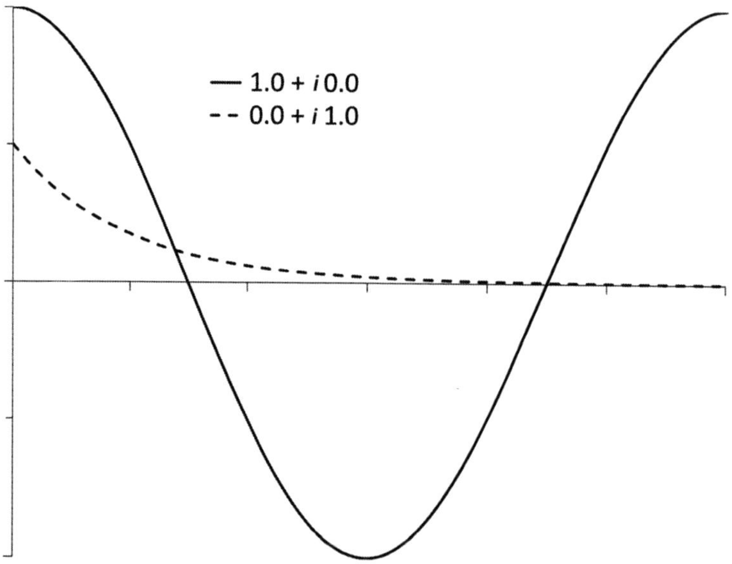

Figure 19.1 Exponential decay by free-electron current. The solid line is the external electric field; the dashed line is the internal electric field. The electric field of the free-electron current completely cancels the external electric field within a spatial wavelength.

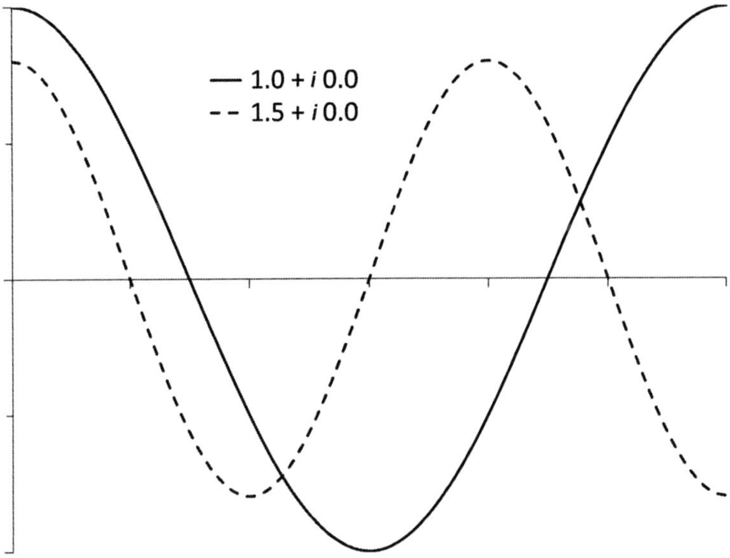

Figure 19.2 Shortened wavelength by bound-electron current. The solid line is the external electric field; the dashed line is the internal electric field. The electric field of the bound-electron current partially cancels the external electric field. The partial cancellation creates a shorter internal wavelength.

The complex refractive indices of free and bound carriers are well documented.[5] The refractive index of free carriers is expressed as a Drude model:[5]

$$N_F^2 = 1 + \left(\frac{\omega_{PF}^2}{-\omega^2 - i\omega/\tau_F} \right) = \varepsilon_R, \tag{19.7}$$

where ω_{PF} is the plasma frequency of free carriers, τ_F is the lifetime of a free carrier, and ε_R is the dielectric constant. The plasma frequency of free carriers may be expressed as

$$\omega_{PF}^2 = \frac{\rho_F q^2}{m\varepsilon_0}, \tag{19.8}$$

where ρ_F is the concentration of free carriers, q is the charge of the carrier, and m is the mass of the carrier.

The corresponding refractive index of bound carriers is expressed as a Lorentz oscillator:[5]

$$N_B^2 = 1 + \left(\frac{\omega_{PB}^2}{-\omega^2 - i\omega/\tau_B + \omega_{RB}^2} \right), \tag{19.9}$$

where ω_{PD} is the plasma frequency of the bound carriers, ω_{RB} is the resonant frequency of the bound state, and τ_B is the lifetime of the bound state. The plasma frequency of bound carriers may be expressed as

$$\omega_{PB}^2 = \frac{\rho_B q^2}{m\varepsilon_0}, \tag{19.10}$$

where ρ_B is the concentration of bound carriers.

If the temporal frequency is much faster than the lifetime τ_B, then the refractive index of a glass becomes

$$N_B^2 = 1 + \left(\frac{\omega_{PB}^2}{\omega_B^2 - \omega^2} \right) = \varepsilon_R; \quad \omega >> \frac{1}{\tau_F}. \tag{19.11}$$

This approximation is the foundation of the Sellmeier formula for refractive index [Eq. (17.1)].

Aluminum represents a combination of Drude and Lorentz spectra. Its refractive index and extinction coefficient are 1.15 and 7.15 at the d line, respectively. Aluminum behaves much like silver; however, it also has a bound oscillator at 800 nm. Consequently, there is a small dip in reflectance at 800 nm. The reflectance of aluminum is near 92% at the d line.

19.3 Relative Partial Dispersion

A SCHOTT glass map frequently plots partial dispersion P_{gF} versus Abbe number. The SCHOTT glass company defined two glass types as normal in dispersion: a crown K7 and a flint F2.

The normal glass types, K7 and F2, define a linear relationship for normal partial dispersion:[52]

$$P_{gF} = -(0.6438) + (0.001682)V_d. \tag{19.12}$$

A departure from normal dispersion is described as relative partial dispersion ΔP_{gF}. It represents the vertical distance from the plot of normal partial dispersion [Eq. (19.12)]. A positive ΔP_{gF} indicates a larger partial dispersion in the blue, while a negative ΔP_{gF} indicates a smaller partial dispersion in the blue. Relative partial dispersion is an important metric for color correction.

19.4 Emission

A dipole moment **p** defines spatial distribution of charge as

$$\mathbf{p} = \int \rho \, \mathbf{r} \, \partial V \,, \tag{19.13}$$

where ρ is the charge density, **r** is the spatial position, and ∂V is the differential volume. The electric potential of the dipole is

$$\Phi = \left(\frac{1}{4\pi\varepsilon_0} \right) \left(\frac{\mathbf{p} \cdot \hat{\mathbf{r}}}{r^2} \right), \tag{19.14}$$

where $\hat{\mathbf{r}}$ is a unit vector. As the distance r grows, the electric field of a radiating dipole becomes

$$\mathbf{E} = \left(\frac{1}{4\pi\varepsilon_0 c^2} \right) \left[\frac{(\hat{\mathbf{r}} \cdot \ddot{\mathbf{p}} \cdot \hat{\mathbf{r}})}{r} \right] \exp(ikr), \tag{19.15}$$

where $\ddot{\mathbf{p}}$ is the temporal curvature of the dipole moment. Exercise of the cross products reveals

$$|\mathbf{E}| = \left(\frac{|\ddot{\mathbf{p}}|}{4\pi\varepsilon_0 c^2} \right) \left(\frac{\sin\theta}{r} \right) \exp(ikr), \tag{19.16}$$

where θ is the angle between $\hat{\mathbf{r}}$ and $\ddot{\mathbf{p}}$. The irradiance of the dipole emission is

$$I_D = \left(\frac{|\ddot{\mathbf{p}}|^2}{32\,\pi^2 \varepsilon_0 c^3} \right) \left(\frac{\sin^2\theta}{r^2} \right). \tag{19.17}$$

19.5 Coherence

The axial version of the space-angle product is[1]

$$\Delta z \, \Delta k_z = 8 \,. \tag{19.18}$$

This may be converted to wavelength as[1]

$$\Delta z \, \Delta\lambda = \frac{4}{\pi} \lambda_P^2 \,, \tag{19.19}$$

where $\Delta\lambda$ is the wavelength range and λ_P is the peak wavelength. The coherence length is

$$s_C = \left(\frac{4}{\pi}\right)\left(\frac{\lambda_P^2}{\Delta\lambda}\right). \tag{19.20}$$

The coherent optical length in cycles is

$$\Lambda_C \approx \frac{\lambda_P}{\Delta\lambda}\ \text{cyc}. \tag{19.21}$$

The coherence length within a refractive medium is

$$\Lambda_C \approx \left(\frac{\lambda_P}{\Delta\lambda}\right)\left(\frac{\lambda_P}{n}\right) = \frac{\lambda_P^2}{n\Delta\lambda}. \tag{19.22}$$

The correlation between two points of an electric field is defined as

$$\Gamma(\mathbf{r}_1,t_1,\mathbf{r}_2,t_2,) = \mathbf{A}(\mathbf{r}_2,t_2)\,\mathbf{A}^*(\mathbf{r}_2,t_2), \tag{19.23}$$

where **A** is the amplitude of the electric field and **A*** is the complex conjugate of **A**. If the correlation has a magnitude of unity, then the system is coherent. If the correlation is null, then the system is incoherent. Figure 19.3 displays the axial coherence of a 50-nm spectral filter.

Coherence determines the structure of interference patterns. Partial reflections of an electric-field pattern can rejoin the transmitted electric field. The contrast of the interference pattern is determined by the correlation of the two waves after separation in both space and time. Coherence length becomes more important as dimensions approach the coherence length.

The correlation of a single dipole field is completely coherent, while the correlation between two dipole fields may be either coherent or incoherent. If two dipoles radiate in constant phase, such as during stimulated emission, then the correlation between dipole fields is coherent. However, if two dipoles radiate in random phase, such as during multiple fluorescent events, then the correlation between the dipole fields is incoherent.

Van Cittert and Zernike developed expressions for the partial coherence of an incoherent source as the Fourier transform of the angular extent of the source.[5] In Eq. (10.2) the partial coherence of a circular incoherent source is stated as

$$\Gamma(\rho) = \frac{2J_1(\rho\, k\,\mathrm{NA}_S)}{\rho\, k\,\mathrm{NA}_S},$$

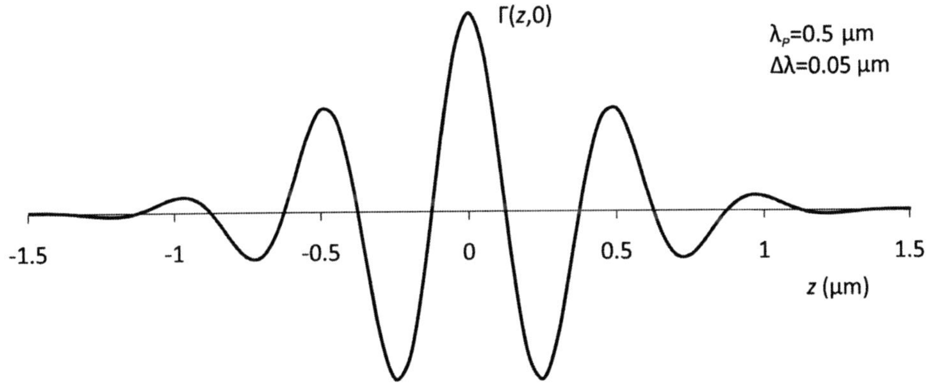

Figure 19.3 Axial coherence of a 50-nm bandpass filter. Peak wavelength of transmission is 500 nm. Bandwidth is 50 nm. Axial coherence is 5 μm by Eq. 19.22.

where ρ is the distance between two points of the illumination field and NA_s is the numerical aperture of the source as defined below:

$$\mathrm{NA}_S = \sin\beta_S\,, \tag{19.24}$$

where β_s is the full angle of the source with respect to the illumination field.

19.6 Gaussian Beam Power

The irradiance profile (power per area) is

$$I = I_0 \exp\left(-2\frac{r^2}{r_G^2}\right). \tag{19.25}$$

The space-angle product of the Gaussian beam is

$$d_G\left(2\,\mathrm{NA}_G\right) = 1.27\lambda\,, \tag{19.26}$$

where NA_G is the Gaussian NA.

The irradiance of a Gaussian profile is

$$I = I_0 \exp\left(\frac{-\pi\,\mathrm{NA}_G^2}{\lambda^2}r^2\right). \tag{19.27}$$

The peak irradiance of a Gaussian beam (power per area) is related to the total power P_0 (quanta per time) by

$$I = 2\frac{\pi \mathrm{NA}_G^2}{\lambda^2} P_0 = \frac{2}{A_G} P_0, \tag{19.28}$$

where P_0 is the power of the beam and A_G is the area of the beam. The power enclosed by a radius is

$$P(r) = P_0\left[1 - \exp\left(-2\frac{r^2}{r_G^2}\right)\right]. \tag{19.29}$$

19.7 Transfer Functions

The amplitude transfer function (ATF) and optical transfer function (OTF) of a lens aperture apply to coherent and incoherent images, respectively.[53] The modulation transfer function (MTF) can apply either the ATF or the OTF.

The MTF of a coherent system applies to the electric field. Thus, the MTF of a coherent system is the ATF. A laser beam is a coherent system, wherein all points have a finite correlation by electric field. The product of consecutive ATFs determines the MTF.

The MTF of an incoherent system applies to the irradiance. Thus, the MTF of an incoherent system is the OTF, which is an autocorrelation of the ATF. Two adjacent fluorophores represent an incoherent system. The phase angle between amplitude point spreads is random over multiple emissions.

The ATF of a circular aperture is

$$\mathrm{ATF}(k) = \begin{Bmatrix} 1 & |k| \le k_M \\ 0 & k_M \le |k| \end{Bmatrix}, \tag{19.30}$$

where k_M is the spatial frequency of the margin in cycles per distance and the ATF cutoff frequency is the spatial frequency of the aperture:

$$k_M = \frac{\mathrm{NA}_M}{\lambda} = \frac{1.22}{\phi_{Airy}}. \tag{19.31}$$

The ATF describes a coherent system where the temporal coherence between points is stationary.

The OTF is the autocorrelation of the ATF. The OTF of a lens with a circular aperture is

$$\mathrm{OTF}(k)=\begin{pmatrix} \dfrac{2}{\pi}\left\{\left[\arccos\left(\dfrac{k}{2k_M}\right)\right]-\left(\dfrac{k}{2k_M}\right)\sqrt{1-\left(\dfrac{k}{2k_M}\right)^2}\right\} & |k|\leq(2k_M) \\ 0 & (2k_M)\leq|k| \end{pmatrix}, \tag{19.32}$$

where the OTF cutoff frequency is

$$(2k_M)=\frac{2\mathrm{NA}_M}{\lambda}=\frac{2.44}{\phi_{Airy}}. \tag{19.33}$$

As displayed in Fig. 19.4, the OTF closely resembles a cone with a height of unity. This cone can be correlated to the Airy transform as follows:

$$\mathrm{OTF}(k)\cong\mathrm{tri}\left(\frac{k}{1.6\,k_M}\right). \tag{19.34}$$

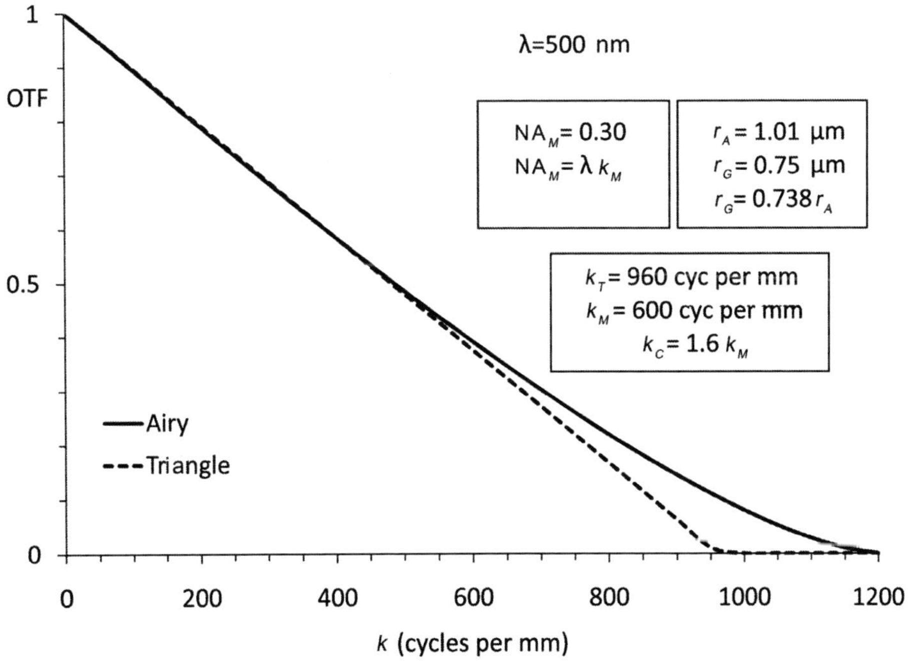

Figure 19.4 MTFs of an Airy pattern and a conical estimate. The solid line is the Airy pattern. The dashed line is the conical estimate. Standard deviation of MTFs is < 0.01. Standard deviation of the corresponding point spread is < 0.03.

The triangle function is defined as

$$\mathrm{tri}(x)=\begin{cases}1-x & (0\le|x|\le 1)\\ 0 & (1<|x|)\end{cases}. \tag{19.35}$$

The marginal frequency is derived from the conical frequency as

$$k_M=\frac{k_C}{1.6}. \tag{19.36}$$

The standard deviation conical estimate from the Airy MTF is < 1% of the expected value up to the marginal frequency. The marginal NA is derived from the conical frequency as

$$\mathrm{NA}_M=\frac{k_C\lambda}{1.6}. \tag{19.37}$$

Equation (19.35) is extremely useful in application to MTF data for an off-the-shelf product. The OTF describes an incoherent system where the temporal coherence between two point spreads is zero.

19.8 Scatter

Surface texture may create a random phase delay during reflection or transmission. Consequently, the exiting electric field $\mathbf{E}_E$ relates to the incident electric field $\mathbf{E}_I$ as

$$\mathbf{E}_E=\mathbf{E}_I\exp\left[i\phi_0\cos\left(k_0x\right)\right], \tag{19.38}$$

where ϕ_0 is the magnitude of the phase delay and k_0 is the spatial frequency of the phase delay. At a small phase delay, this becomes

$$\mathbf{E}_E=\mathbf{E}_I\left[1+i\phi_0\cos\left(k_0x\right)\right]. \tag{19.39}$$

The scatter coefficient for the electric field is

$$s=1-\frac{\mathbf{E}_E}{\mathbf{E}_I}=-i\phi_0\cos\left(k_0x\right). \tag{19.40}$$

The scatter, or scatterance, applies to the irradiance as

$$S = \left\langle |s|^2 \right\rangle = \phi^2 , \qquad (19.41)$$

where ϕ is the rms of the phase delay. The scatterance is frequently called the total integrated scatter (TIS).

There a three specific formats of the rms phase delay. The rms phase delay for transmission is

$$\phi_T = \Delta n \left(\frac{2\pi\sigma}{\lambda} \right) , \qquad (19.42)$$

where σ is the rms surface texture. The rms phase delay for an external reflection is

$$\phi_{ER} = 2 \left(\frac{2\pi\sigma}{\lambda} \right) . \qquad (19.43)$$

The rms phase delay for an internal reflection is

$$\phi_{IR} = 2n \left(\frac{2\pi\sigma}{\lambda} \right) . \qquad (19.44)$$

19.9 Interference Filters

An interference filter comprises a dielectric stack of high- and low-index materials. The repeated structure is similar to a crystalline structure wherein an x ray is diffracted based on a lattice vector. The analogous filter diffraction is displayed in Fig. 19.5, where the reflected wave vector $\mathbf{k}_R$ is the sum of the incident wave vector $\mathbf{k}_{In}$ and the filter wave vector $\mathbf{k}_F$:

$$\mathbf{k}_R = \mathbf{k}_{In} + \mathbf{k}_F . \qquad (19.45)$$

The wave vector of reflection is dependent on the angle of incidence as

$$2\mathbf{k}_R \cdot \mathbf{k}_F = 2k_R k_F \cos\theta = k_F^2 . \qquad (19.46)$$

The spatial frequency of reflection obeys

$$k_R \cos\theta = \frac{1}{2} k_F . \qquad (19.47)$$

At normal incidence, the spatial frequency of reflection is half that of the filter lattice.

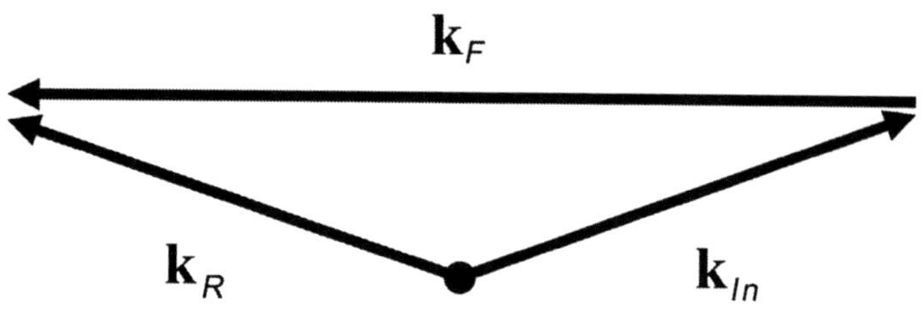

Figure 19.5 Wave vectors of the interference filter. A reflected vector is the sum of the incident vector and the filter vector. Blue shift is indicated by longer wave vectors at a larger angle with the filter vector.

19.10 Shot Noise

The contrast of a fluorophore with a short lifetime is frequently limited by the background shot noise. The quantum of an object with a short lifetime is dependent on an exponential decay

$$Q_{OSL} \propto 1 - \exp\left(\frac{-t}{\tau_F}\right) \quad (t \approx \tau_F), \tag{19.48}$$

where t is the integration time and τ_F is the lifetime of the fluorophore. The quantum of the dark current is linearly dependent on the integration time:

$$Q_{DC} \propto t. \tag{19.49}$$

The quantum of read noise is independent of the integration time:

$$Q_{RN} \propto 1. \tag{19.50}$$

The quantum of the background shot noise is dependent on the square root of integration time:

$$Q_{BSN} \propto t^{1/2}. \tag{19.51}$$

The shot noise of the background may easily become larger than the read noise or the object quantum.

The contrast of a fluorophore with a long lifetime is frequently limited by the shot noise of the quanta. The quantum of a fluorophore with a long lifetime is linearly dependent on the integration time:

$$Q_{OLL} \propto t \quad (t << \tau_F). \tag{19.52}$$

The contrast with dark current is stationary in time, the contrast with read noise grows linearly with time, and the contrast with shot noise grows with the square root of time.

Beyond increasing the integration time, the effects of background and noise may be reduced by optical design. These expressions for contrast provide a foundation for maximization of contrast through selection of NA and magnification.

Appendix: Prescriptions

The prescriptions used in this book are summarized below.

The magnification *M* has several formats: spatial enlargement (10X), spatial enlargement with infinity correction (10XI), spatial reduction (10XR), or angular enlargement (10XA) .

The focal lengths are important to macro- and microlens formats. A microscope objective is normally defined by magnification, numerical aperture, and tube length. The tube length (TL) may specify the focal length of the tube lens for enlargement with infinity correction.

The field of view (FOV) can be spatial (25 mm) or angular (15 deg). An attempt is made to make the field of view equivalent the 12-mm diameter of a typical CCD sensor.

The image convergence is specified by paraxial numerical aperture (NA), image-space numerical aperture (ISNA), object-space numerical aperture (OSNA), *f*-number (*f*/#), or image-space *f*/#, which indicates the *f*/# inflation.

Aspheric parameters indicate the following: conic indicates a conic constant; A4 indicates the scale of radius to 4th power; A6 indicates the scale of radius to 6th power.

Prescription 4.1 Spherical aberration of a spherical lens.
(M = 0, NA = 0.25, f = 20 mm, FOV = 0 deg)

Surf	Radius	Thickness	Glass	Diameter	n_d	ν_d
OBJ	∞	∞		0	-	-
1	∞	20		10	-	-
STO	10.37	30.20	BK7	60	1.517	64.2
IMA	∞			0.524	-	-

Prescription 4.2 Coma in an aspheric lens.
(M = 0, NA = 0.25, f = 20 mm, FOV – 1.0 deg)

Surf	Radius	Thickness	Glass	Diameter	Conic	n_d	ν_d
OBJ	∞	∞		0.00		-	-
STO	∞	0.00		10.00		-	-
2.00	10.34	30.33	BK7	12.00	–0.4337	1.517	64.2
IMA	∞			0.80		-	-

Prescription 4.3 Astigmatism and lateral color in a spherical lens.
(M = 0, NA = 0.02, f = 20 mm, FOV = 15 deg)

Surf	Radius	Thickness	Glass	Diameter	n_d	ν_d
OBJ	∞	∞		0	-	-
STO	∞	15.00		0.7	-	-
2	10.34	22.19	BK7	12	1.517	64.2
IMA	∞			0.8	-	-

Prescription 4.4 Axial color in an aspheric lens.
(M = 0, NA = 0.25, f = 20 mm, FOV = 0 deg)

Surf	Radius	Thickness	Glass	Diameter	Conic	n_d	ν_d
OBJ	∞	∞		0.00	0	-	-
1	∞	20.00		10.00	0	-	-
STO	10.34	30.34	BK7	20.00	–0.4334	1.517	64.2
IMA	∞			0.07	0	-	-

Prescriptions 7.1–7.8 5X doublet from OptoSigma, Part No. 026-0270.
(M = 5XI, TL = 200 mm, NA = 0.12, f = 40 mm, FOV = 2.4 mm)

Surf	Radius	Thickness	Glass	Diameter	Comment	n_d	ν_d
OBJ	∞	∞		0.00		-	-
STO	23.40	4.10	BK7	12.70	026-0270	1.517	64.2
2.00	–17.60	2.00	SF2	12.70		1.648	33.8
3.00	–60.31	36.84		12.70		-	-
IMA	∞			2.412318		-	-

Prescription 9.1 10XR double gauss.[12]
(M = 10XR, $f/\#$ = 1.4, Image-space $f/\#$ = 1.5, f = 52 mm, FOV = 12 mm)

Surf	Radius	Thickness	Glass	Diameter	n_d	ν_d
OBJ	∞	530.86		120.8	-	-
1	∞	10.00		15.0	-	-
2	79.02	5.43	LaFN21	44.0	1.788	47.5
3	–387.15	0.25		44.0	-	-
4	32.41	4.57	LaFN21	38.0	1.788	47.5
5	62.73	3.63		36.0	-	-
6	∞	2.30	SF15	34.0	1.699	30.1
7	27.90	5.17		30.0	-	-
STO	∞	5.97		27.0	-	-
9	–23.48	2.30	SF15	30.0	1.699	30.1
10	–650.00	6.57	LaFN21	34.0	1.788	47.5
11	–32.30	0.25		36.0	-	-
12	–218.50	3.56	LaFN21	38.0	1.788	47.5
13	–58.10	0.25		38.0	-	-
14	128.60	3.64	LaFN21	38.0	1.788	47.5
15	–175.70	43.05		38.0	-	-
IMA	∞			12.3	-	-

Prescription 9.2 10XR microlens, from Matsui.[13]
(M = 10XR, $f/\#$ = 2.9, Image-space $f/\#$ = 3.2, f = 105 mm, FOV = 12 mm)

Surf	Radius	Thickness	Glass	Diameter	n_d	ν_d
OBJ	∞	1094.38		120.0	-	-
1	∞	–20.00		39.6	-	-
2	∞	6.00	Silica	50.0	1.458	67.8
3	∞	14.00		50.0	-	-
4	88.80	4.30	E-LaK13	44.0	1.694	53.2
5	–4656.70	7.70		39.1	-	-
6	33.12	6.00	E-LaK14	36.0	1.697	55.5
7	85.73	0.70		32.0	-	-
8	169.57	2.00	E-SF2	34.0	1.648	33.8
9	27.68	23.48		30.0	-	-
10	∞	–3.00		24.1	-	-
STO	∞	3.00		24.0	-	-
12	–29.10	2.00	E-F2	24.0	1.620	36.3
13	–420.00	7.50	E-LaK13	27.0	1.694	53.2
14	–42.41	0.10		30.0	-	-
15	258.00	4.30	E-LaK13	31.0	1.694	53.2
16	–65.23	8.35		31.0	-	-
17	–92.06	3.50	E-SF6	31.0	1.805	25.4
18	–56.03	12.96		31.0	-	-
19	–41.90	2.00	E-LaSF013	26.0	1.804	39.6
20	83.22	1.00		30.0	-	-
21	105.00	5.00	E-LaF04	32.0	1.757	47.8
22	–89.47	44.50		32.0	-	-
IMA	∞			12.0	-	-

Prescription 9.3 2XR microlens, from Matsui (1998).[13]
(M = 2XR, f/# = 2.9, Image-space $f/\#$ = 4.3, f = 105 mm, FOV = 12 mm)

Surf	Radius	Thickness	Glass	Diameter	n_d	ν_d
OBJ	∞	240.23		23.0	-	-
1	∞	–20.00		32.3	-	-
2	∞	6.00	Silica	50.0	1.458	67.8
3	∞	14.00		50.0	-	-
4	88.80	4.30	E-LaK13	44.0	1.694	53.2
5	–4656.70	7.70		32.0	-	-
6	33.12	6.00	E-LaK14	36.0	1.697	55.5
7	85.73	0.70		32.0	-	-
8	169.57	2.00	E-SF2	34.0	1.648	33.8
9	27.68	16.93		30.0	-	-
10	∞	–3.00		24.3	-	-
STO	∞	3.00		24.0	-	-
12	–29.10	2.00	E-F2	24.0	1.620	36.3
13	–420.00	7.50	E-LaK13	27.0	1.694	53.2
14	–42.41	0.10		30.0	-	-
15	258.00	4.30	E-LaK13	31.0	1.694	53.2
16	–65.23	35.80		31.0	-	-
17	–92.06	3.50	E-SF6	31.0	1.805	25.4
18	–56.03	12.96		31.0	-	-
19	–41.90	2.00	E-LaSF013	26.0	1.804	39.6
20	83.22	1.00		30.0	-	-
21	105.00	5.00	E-LaF04	32.0	1.757	47.8
22	–89.47	44.50		32.0	-	-
IMA	∞			12.0	-	-

Prescription 9.4 1X microlens, from Matsui (1998).[13]
(M = 1X, $f/\#$ = 2.9, Image-space $f/\#$ = 5.5, f = 105 mm, FOV = 12 mm)

Surf	Radius	Thickness	Glass	Diameter	n_d	ν_d
OBJ	∞	149.72		12.1	-	-
1	∞	–20.00		29.2	-	-
2	∞	6.00	Silica	50.0	1.458	67.8
3	∞	14.00		50.0	-	-
4	88.80	4.30	E-LaK13	44.0	1.694	53.2
5	–4656.70	7.70		28.9	-	-
6	33.12	6.00	E-LaK14	36.0	1.697	55.5
7	85.73	0.70		32.0	-	-
8	169.57	2.00	E-SF2	34.0	1.648	33.8
9	27.68	11.76		30.0	-	-
10	∞	–3.00		24.5	-	-
STO	∞	3.00		24.0	-	-
12	–29.10	2.00	E-F2	24.1	1.620	36.3
13	–420.00	7.50	E-LaK13	27.0	1.694	53.2
14	–42.41	0.10		30.0	-	-
15	258.00	4.30	E-LaK13	31.0	1.694	53.2
16	–65.23	64.69		31.0	-	-
17	–92.06	3.50	E-SF6	31.0	1.805	25.4
18	–56.03	12.96		31.0	-	-
19	–41.90	2.00	E-LaSF013	26.0	1.804	39.6
20	83.22	1.00		30.0	-	-
21	105.00	5.00	E-LaF04	32.0	1.757	47.8
22	–89.47	44.50		32.0	-	-
IMA	∞			12.0	-	-

Prescription 9.5 2XR TSL, prescription patent pending by Seward (2008).[14]
(M = 2XR, $f/\#$ = 3.5, Image-space $f/\#$ = 3.9, f = 193 mm, FOV = 12 mm)
(Glass types by Ohara Inc: lanthanum with low-index LAL, titanium with medium-index TIM, lanthanum with medium-index LAM.)

Surf	Radius	Thickness	Glass	Diameter	Conic	A4	A6	n_d	ν_d
OBJ	Infinity	414.12		24.44				-	-
1	Infinity	11.80	Silica	60.00	0	1.22×10^{-08}		1.458	88.3
2	–249.50	15.80		60.00				-	-
STO	Infinity	-		54.00				-	-
4	Infinity	6.00	Silica	60.00				1.458	88.3
5	Infinity	15.80		60.00				-	-
6	161.70	15.60	S-LaL7	60.00				1.652	76.7
7	–161.70	10.00	S-TIH4	60.00				1.755	36.7
8	Infinity	281.60		60.00				-	-
9	21.00	11.40	S-LaL7	30.00	0	-1.52×10^{-06}	-4.58×10^{-09}	1.652	76.7
10	Infinity	6.60	S-LaM7	22.00				1.749	46.8
11	18.30	11.40		16.00				-	-
12	Infinity	1.00	Silica	14.00				1.458	88.3
13	Infinity	1.00		14.00				-	-
IMA	Infinity			12.02				-	-

Prescription 10.6 Aspheric plus singlet. Production parts from OptoSigma.
(M = 4XR, ISNA = 0.25, f = 22 mm, FOV = 1.2 mm)

Surf	Radius	Thickness	Glass	Diameter	Conic	Comment	n_d	ν_d
OBJ	∞	93.12		4.9	0		-	-
STO	∞	11.54		13.0	0	LS	-	-
2	∞	4.20	BK7	30.0	0	011-2280	1.517	64.2
3	–51.90	5.00		30.0	0		-	-
4	13.90	11.00	B270	30.0	–0.544	023-2390	1.523	58.6
5	∞	18.21		11.3	0		-	-
6	∞	1.00	BK7	10.0	0	Slide	1.517	64.2
7	∞	20.00		1.3	0		-	-
8	∞	–20.00		12.2	0	VLS	-	-
IMA	∞			1.3	0		-	-

Prescription 10.7 Aspheric plus doublet. Custom doublet, production aspheric from OptoSigma. (M = 4XR, ISNA = 0.25, f = 19 mm, FOV = 1.2 mm)

Surf	Radius	Thickness	Glass	Diameter	Conic	Comment	n_d	ν_d
OBJ	∞	72.95		4.8	0		-	-
STO	∞	7.11		13.0	0	LS	-	-
2	–25.82	3.00	SF2	20.0	0	Custom	1.648	33.8
3	16.27	10.00	BK7	20.0	0	Custom	1.517	64.2
4	–16.27	5.00		20.0	0		-	-
5	13.90	11.00	B270	30.0	–0.5443	023-2390	1.523	58.6
6	∞	19.95		12.1	0		-	-
7	∞	1.00	BK7	10.0	0	Slide	1.517	64.2
8	∞	20.00		1.3	0		-	-
9	∞	–20.00		11.4	0	VLS	-	-
IMA	∞			1.3	0		-	-

Prescription 10.8 Abbe illumination. Production parts from OptoSigma. (M = 4XR, ISNA = 0.25, f = 19 mm, FOV = 1.2 mm)

Surf	Radius	Thickness	Glass	Diameter	Comment	n_d	ν_d
OBJ	∞	54.05		4.8		-	-
1	10.38	5.20	BK7	15.0	011-1180	1.517	64.2
2	∞	5.00		6.8		-	-
STO	∞	9.87		5.6		-	-
4	5.19	3.40	BK7	8.0	011-0470	1.517	64.2
5	∞	2.50		8.0		-	-
6	∞	1.00	BK7	10.0	Slide	1.517	64.2
7	∞	20.00		1.4		-	-
8	∞	–20.00		11.5		-	-
IMA	∞			1.4		-	-

Prescription 10.9 Aspheric Abbe. Two custom aspheric elements of B270. Conic constant is the only aspheric parameter. (M = 4XR, ISNA = 0.25, f = 18 mm, FOV = 1.2 mm)

Surf	Radius	Thickness	Glass	Diameter	Conic	Comment	n_d	ν_d
OBJ	∞	56.01		4.8			-	-
1	10.38	5.20	B270	15.0	–0.852	Low Tg	1.523	58.6
2	∞	5.00		7.1			-	-
STO	∞	10.45		5.6			-	-
4	5.19	3.40	B270	8.0	–0.633	Low Tg	1.523	58.6
5	∞	2.50		8.0			-	-
6	∞	1.00	BK7	10.0	slide		1.517	64.2
7	∞	20.00		1.2		VLS	-	-
8	∞	–20.00		11.7			-	-
IMA	∞			1.2			-	-

Prescription 10.10 TIRF illumination, derived from U.S. Patent No. 6,519,092 by Yamaguchi.[23] (M = 60XI, TL = 200 mm, ISNA = 1.0, f = 3.3 mm, FOV = 60 deg)

Surf	**Radius**	**Thickness**	**Glass**	**Diameter**	**Dec Y**		
OBJ	∞	–15.81		0.0		n_d	ν_d
STO	∞	0.00		1.6		-	-
2	-	0.00		-	4.22	1.000	-
3	∞	–4.90		30.0		-	-
4	∞	1.00		25.0		-	-
5	∞	1.00		12.0		-	-
6	8.17	3.00	E-SF6	10.4		1.805	25.4
7	204.67	2.60	E-LaSF013	10.4		1.804	39.6
8	4.80	1.70		7.1		-	-
9	∞	1.70		6.6		-	-
10	–4.82	5.00	E-LaSF013	7.1		1.804	39.6
11	238.40	5.20	N-PSK58	15.0		1.569	71.2
12	–8.78	0.15		15.0		-	-
13	–25.25	1.00	KF9	16.4		1.523	51.5
14	48.48	6.30	CaF_2	19.0		1.434	95.0
15	–11.82	1.60	KzFH1	19.0		1.613	44.4
16	–18.67	0.10		21.0		-	-
17	28.66	8.00	CaF_2	22.0		1.434	95.0
18	–14.80	1.10	KzFH1	22.0		1.613	44.4
19	–47.67	0.15		24.0		-	-
20	16.27	9.40	CaF_2	23.6		1.434	95.0
21	–20.05	1.00	KzFH1	23.6		1.613	44.4
22	84.75	0.15		23.2		-	-
23	13.453	6.8	N-PSK58	20.8		1.569	71.2
24	–34.23	1.00	F5	20.8		1.603	38.1
25	27.891	0.10		20.4		-	-
26	7.247	3.75	N-PSK58	13.6		1.569	71.2
27	13.716	0.10		12.4		-	-
28	3.716	3.60	LaSF35	7.4		2.022	29.1
29	1.332	0.65	TYPE A	2.3		1.515	41.6
30	∞	0.14	TYPE A	6		1.515	41.6
31	∞	0.17	D-263-C	10.000		1.523	54.6
IMA	∞		D-263-C	0.352		1.523	54.6

Prescription 11.1 Perfect 10X objective in air.
[M = 10XI, ISNA = 0.25, f = 20.0 mm, FOV = 1.2 mm,1-nm defocus (1×10^{-6})]

Surf	Radius	Thickness	Glass	Diameter	Comment	n_d	ν_d
OBJ	∞	–20.00		1.2	Virtual object	-	-
STO	20.00	20.00		10.0	Lens stop	-	-
2	∞	0.00	D263M	10.0	Cover	1.523	52.3
IMA	∞			1.2	Image	-	-

Prescription 11.2 10X objective with cover glass in place of air.
(M = 10XI, TL = 200 mm, ISNA = 0.25, f = 20.0 mm, FOV = 1.2 mm)

Surf	Radius	Thickness	Glass	Diameter	Comment	n_d	ν_d
OBJ	∞	–20.00		1.2	Virtual object	-	-
STO	20.00	19.89		10.0	Lens stop	-	-
2	∞	0.17	D263M	10.0	Cover	1.523	52.3
IMA	∞		D263M	1.2	Image	-	-

Prescription 11.3 10X objective with microscope slide in place of air.
(M = 10XI, TL = 200 mm, ISNA = 0.25, f = 20.0 mm, FOV = 1.2 mm)

Surf	Radius	Thickness	Glass	Diameter	Comment	n_d	ν_d
OBJ	∞	–20.00		1.2	Virtual object	-	-
STO	20.00	19.35		10.0	Lens stop	-	-
2	∞	1.00	D263M	10.0	Cover	1.523	52.3
IMA	∞		D263M	1.2	Image	-	-

Prescription 11.4 40X objective with silica cover in place of glass.
(M = 40XI, TL = 200 mm, ISNA = 0.75, f = 5.0 mm, FOV = 0.3 mm)

Surf	Radius	Thickness	Glass	Diameter	Comment	n_d	ν_d
OBJ	∞	–0.17	D263M	0.3	Virtual object	1.523	52.3
1	∞	–4.89		0.5		-	-
STO	5.00	4.88		8.6	Lens stop	-	-
3	∞	0.17	SILICA	10.0	Cover	1.458	67.8
IMA	∞		SILICA	0.3	Image	1.458	67.8

Prescription 11.5 40X objective with tilted cover glass.
(ISNA = 0.75, f = 5.0 mm)

Surf	Radius	Thickness	Glass	Diameter	Comment	Tilt	n_d	ν_d
OBJ	∞	–0.17	D263M	0.3	Virtual object		1.523	52.3
1	∞	–4.89		0.5			-	-
STO	5.00	4.89		8.6	Lens stop		-	-
3	-	0.00		-		0.50	-	-
4	∞	0.17	D263M	10.0	Cover		1.523	52.3
5	-	0.00		-		–1.00	-	-
IMA	∞		D263M	10.0	Image		1.523	52.3

Prescription 11.6 60X objective with silica cover in place of glass. Oil must vary in thickness for compensation of silica in place of glass.
(M = 60XI, TL = 200 mm, ISNA = 1.4, f = 3.3 mm, FOV = 0.2 mm)

Surf	Radius	Thickness	Glass	Diameter	Comment	n_d	ν_d
OBJ	∞	–0.17	D263M	0.2	Virtual object	1.523	52.3
1	∞	–4.88	TYPE A	1.0	Oil	1.515	41.6
STO	5.00	4.87	TYPE A	9.3	Lens stop	1.515	41.6
3	∞	0.17	Silica	10.0	Cover	1.458	67.8
IMA	∞		Silica	0.6	Image	1.458	67.8

Prescription 12.1 Aplanatic front end of the scaled-divergence kind.
(M = 2X, ISNA = 0.75, f = 8.2 mm, FOV = 0.0 mm)

Surf	Radius	Thickness	Glass	Diameter	Conic	Comment	n_d	ν_d
OBJ	∞	–0.67		0.0	0		-	-
STO	∞	–9.95		0.5	0		-	-
2	4.22	6.00	N-BK7	7.0	0	S1	1.517	-
3	∞	1.00	N-BK7	4.0	0	Oil	1.517	-
IMA	–7.94		N-BK7	4.0	0		1.517	-

Prescription 12.2 Aplanatic front end with three aplanatic surfaces.
(M = 3.5X, ISNA = 1.00, f = 8.8 mm, FOV = 0.0 mm)

Surf	Radius	Thickness	Glass	Diameter	Conic	Comment	n_d	ν_d
OBJ	∞	–0.71		0.0	0		-	-
STO	∞	–24.50		0.5	0		-	-
2	10.01	5.00	N-BK7	16.7	0	S3	1.517	-
3	11.62	1.00		13.2	0	S2	-	-
4	4.22	6.00	N-BK7	8.4	0	S1	1.517	-
5	∞	1.00	N-BK7	4.0	0	Oil	1.517	-
IMA	–4.47		N-BK7	4.0	0		1.517	-

Prescription 12.3 10X plan achromat. Derived from Fukatake,[21] U.S. Patent No. 6,128,139. (M = 10XI, TL = 200 mm, NA = 0.25, f = 20 mm, FOV = 1.2 mm)

Surf	Radius	Thickness	Glass	Diameter	Comment	n_d	ν_d
OBJ	∞	∞		0.0		-	-
1	∞	13.38		10.2	Flange	-	-
STO	∞	6.20		11.4		-	-
3	19.30	3.10	SSK1	12.0		1.617	54.0
4	–19.30	1.85	E-LaSF013	12.0		1.804	39.6
5	53.53	17.45		11.0		-	-
6	42.00	2.55	E-LaSF015	12.0		1.804	46.6
7	–42.00	0.60		12.0		-	-
8	11.54	5.35	E-SSK5	10.0		1.658	50.9
9	–20.10	2.95	SF6	10.0		1.805	25.4
10	11.04	6.57		8.0		-	-
IMA	∞			1.2		-	-

Prescription 12.4 40X fluor, derived from Misawa,[22] U.S. Patent No. 5,699,196. (M = 40XI, TL = 200 mm, NA = 0.75, f = 5 mm, FOV 0.3 mm)

Surf	Radius	Thickness	Glass	Diameter	n_d	ν_d
OBJ	∞	∞		0.0	-	-
1	∞	5.00		10.0	-	-
STO	–8.84	1.80	E-FK5	10.0	1.487	70.4
3	10.94	4.76	LLF6	13.0	1.532	48.8
4	–16.02	30.00		13.0	-	-
5	27.49	1.50	F5	13.0	1.603	38.0
6	12.55	5.60	N-FK56	13.0	1.434	95.0
7	–11.24	1.50	SF8	13.0	1.689	31.2
8	–39.44	0.23		13.0	-	-
9	29.31	3.61	E-FKH1	13.0	1.498	82.5
10	–29.75	1.58		13.0	-	-
11	6.72	5.48	N-FK56	11.0	1.434	95.0
12	–16.61	1.00	K3	11.0	1.518	59.0
13	15.13	0.18		8.0	-	-
14	3.36	3.33	N-LaF34	6.6	1.773	49.6
15	2.25	1.09		3.0	-	-
16	∞	0.17	D-263-C	3.0	1.523	54.6
IMA	∞			0.4	-	-

Prescriptions 12.5–12.7 60X TIRF, derived from Yamaguchi,[23] U.S. Patent No. 6,519,092. (M = 60XI, TL = 200 mm, ISNA = 1.4, f = 3.3 mm, FOV = 0.2 mm)

Surf	Radius	Thickness	Glass	Diameter	Comment	n_d	ν_d
OBJ	∞	∞		0.0		-	-
1	∞	–4.90		30.0	Flange	-	-
2	∞	1.00		25.0	Barrel	-	-
3	∞	1.00		12.0	Lens stop	-	-
4	8.17	3.00	E-SF6	10.4	L8	1.805	25.4
5	204.67	2.60	E-LaSF013	10.4		1.804	39.6
6	4.80	1.70		7.1		-	-
STO	∞	1.70		6.6		-	-
8	–4.82	5.00	E-LaSF013	7.1	L7	1.804	39.6
9	238.40	5.20	N-PSK58	15.0		1.569	71.2
10	–8.78	0.15		15.0		-	-
11	–25.25	1.00	KF9	16.4	L6	1.523	51.5
12	48.48	6.30	CaF_2	19.0		1.434	95.0
13	–11.82	1.60	KzFH1	19.0		1.613	44.4
14	–18.67	0.10		21.0		-	-
15	28.66	8.00	CaF_2	22.0	L5	1.434	95.0
16	–14.80	1.10	KzFH1	22.0		1.613	44.4
17	–47.67	0.15		24.0		-	-
18	16.27	9.40	CaF_2	23.6	L4	1.434	95.0
19	–20.05	1.00	KzFH1	23.6		1.613	44.4
20	84.75	0.15		23.2		-	-
21	13.45	6.80	N-PSK58	20.8	L3	1.569	71.2
22	–34.23	1.00	F5	20.8		1.603	38.1
23	27.891	0.10		20.4		-	-
24	7.247	3.75	N-PSK58	13.6	L2	1.569	71.2
25	13.716	0.10		12.4		-	-
26	3.716	3.60	LaSF35	7.4	L1	2.022	29.1
27	1.332	0.65	K3	2.3		1.518	58.9
28	∞	0.141505	K3	6.0	Oil	1.518	58.9
29	∞	0.17	D-263-C	10.0	Coverslip	-	-
IMA	∞		D-263-C	23.37486		1.523	54.6

Prescriptions 12.9–12.10 Aplanat, from Smith.[24]
(M = 100X, TL = 180 mm, ISNA = 1.3, f = 1.8 mm, FOV = 0.12 mm)

Surf	Radius	Thickness	Glass	Diameter	n_d	ν_d
OBJ	∞	180.00		12.1	-	-
STO	4.68	1.70	CaF_2	4.6	1.434	95.0
2	–3.86	1.04	F2	4.6	1.620	36.4
3	41.30	0.09		4.3	-	-
4	3.51	1.36	CaF_2	4.2	1.434	95.0
5	–5.10	0.66	BaF2	4.2	1.570	49.4
6	–310.90	0.24		4.2	-	-
7	1.93	1.08	BK1	3.4	1.510	63.5
8	7.71	0.02		3.0	-	-
9	0.77	0.93	BK7	1.5	1.517	64.2
10	∞	0.10	TYPE A	1.3	1.515	41.6
11	∞	0.17	K5	1.3	1.522	59.5
IMA	∞		K5	0.1	1.522	59.5

Prescription 12.11 Schwarzschild, after geometric relationships in Ref. 25.
(M = 10XI, TL = 200 mm, ISNA = 0.25, f = 20, FOV = 10 mm)

Surf	Radius	Thickness	Glass	Diameter	n_d	ν_d
OBJ	∞	∞		0.0	-	-
1	∞	10.00		13.0	-	-
2	∞	40.00		10.0	-	-
STO	24.72	–40.00	Mirror	10.0	1.00	-
4	64.72	40.00	Mirror	44.4	1.00	-
5	∞	44.72		25.1	-	-
6	∞	0.00		1.2	-	-
IMA	–20.00			10.00	-	-

Prescription 12.12 Internal parabola, after Larson et al.[26] and Krogmeier et al.[27]
(M = 20X, TL = 200 mm, ISNA = 1.0, f = 10, FOV = 0.6 mm)

Surf	Radius	Thickness	Glass	Diameter	Conic constant	n_d	ν_d
OBJ	∞	∞		0.0		-	-
1	∞	10.00		0.0		-	-
2	∞	1.00	Silica	40.0		1.458	67.8
3	∞	13.63	Silica	40.0		1.458	67.8
STO	–29.26	–13.63	Mirror	40.0	–1.000	1.458	67.8
5	∞	–1.00	Silica	3.3		1.458	67.8
IMA	∞		Silica	0.7		1.458	67.8

Prescription 13.1 Production doublet as a tube lens, from Linos Photonics No. 322271. (M = 1X, TL = 200 mm, f = 200 mm, ISNA = 0.025, FOV = 12 mm)

Surf	Radius	Thickness	Glass	Diameter	Comment	n_d	ν_d
OBJ	∞	∞		0.0		-	-
STO	∞	200.00		5.0	L-stop	-	-
2	92.39	5.30	N-BK7	31.5	322271	1.517	64.2
3	–84.75	2.80	F4	31.5		1.617	36.6
4	–1074.60	0.00		31.5		-	-
5	∞	194.71		27.0		-	-
IMA	∞			21.9	Sense	-	-

Prescription 13.2 Custom doublet as a tube lens. (M = 1X, TL = 200 mm, ISNA = 0.025, f = 200 mm, FOV = 12 mm)

Surf	Radius	Thickness	Glass	Diameter	Comment	n_d	ν_d
OBJ	∞	∞		0.0		-	-
STO	∞	200.00		10.0	L-stop	-	-
2	131.27	8.00	N-BK7	31.5	322271	1.517	64.2
3	–67.26	5.00	F4	31.5		1.617	36.6
4	–238.12	194.31		31.5		-	-
IMA	∞			12.0	Sense	-	-

Prescriptions 13.3–13.4 Doublet-pair tube lens, from Misawa,[22] U.S. Patent No. 5,699,196. (M = 1X, TL = 200 mm, ISNA = 0.025, f = 200 mm, FOV = 12 mm) Required 5X scale of Table 2.

Surf	Radius	Thickness	Glass	Diameter	n_d	ν_d
OBJ	∞	∞		0.0	-	-
STO	∞	140.00		10.0	-	-
2	75.05	5.10	<MODEL>	30.0	1.623	57.1
3	–75.05	2.00	<MODEL>	30.0	1.750	35.2
4	1600.73	7.50		30.0	-	-
5	50.26	5.10	<MODEL>	30.0	1.668	42.0
6	–84.55	1.80	<MODEL>	30.0	1.613	44.4
7	36.91	168.43		30.0	-	-
IMA	∞			12.0	-	-

Prescription 13.6 Doublet-pair tube lens with real glass, from Misawa,[22] U.S. Patent No. 5,699,196. Real glass types of the 1990s are not available in 2009.
(M = 1X, TL = 200 mm, ISNA = 0.025, f = 200 mm, FOV = 12 mm)

Surf	Radius	Thickness	Glass	Diameter	n_d	ν_d
OBJ	∞	∞		0.0	-	-
STO	∞	140.00		10.0	-	-
2	75.05	5.10	SK10	30.0	1.623	56.9
3	–75.05	2.00	LaFN7	30.0	1.750	35.0
4	1600.73	7.50		30.0	-	-
5	50.26	5.10	BaSF6	30.0	1.668	41.9
6	–84.55	1.80	KzFSN4	30.0	1.613	44.3
7	36.91	169.25		30.0	-	-
IMA	∞			12.0	-	-

Prescription 13.9 Cube within a finite conjugate. (Strehl ratio at the d line is 0.9.)
(M = 1X, TL = 200 mm, ISNA = 0.025, FOV = 12 mm)

Surf	Radius	Thickness	Glass	Diameter	n_d	ν_d
OBJ	∞	–200		12.0	-	-
STO	∞	150.00		10.0	-	-
2	∞	50.00	Silica	50.0	1.458	67.8
3	∞	15.72		11.8	-	-
IMA	∞			12.0	-	-

Prescription 13.10 Tilted plate within a finite conjugate. (Strehl ratio at the d line is 0.9.)
(M = 1X, TL = 200 mm, ISNA = 0.025, FOV = 12 mm)

Surf	Radius	Thickness	Glass	Diameter	Tilt	n_d	ν_d
OBJ	∞	–200		12.0		-	-
STO	∞	150.00		10.0		-	-
2	-	0.00		-	–45	-	-
3	∞	1.38	Silica	50.0		1.458	67.8
4	∞	–1.38		50.0		-	-
5	-	50.42		-	45	-	-
IMA	∞			12.9		-	-

Prescription 13.11 Thick plate warped within ∞ correction. (Strehl ratio at the d line is 0.9.)
(M = 1X, TL = 200 mm, ISNA = 0.00, FOV = 0 deg)

Surf	Radius	Thickness	Glass	Diameter	Tilt	n_d	ν_d
OBJ	∞	∞		0.0		-	-
STO	∞	150.00		10.0		-	-
2	-	0.00		-	–45	-	-
3	–580.62	3.00	Silica	50.0		1.458	67.8
4	–583.62	–3.00		50.0		-	-
5	-	50.00		-	45	-	-
IMA	∞			0.0		-	-

Prescription 13.12 Thin plate warped within ∞ correction. (Strehl ratio at the d line is 0.9.)
(M = 1X, TL = 200 mm, ISNA = 0.00, FOV = 0 deg)

Surf	Radius	Thickness	Glass	Diameter	Tilt	n_d	ν_d
OBJ	∞	∞		0.0		-	-
STO	∞	150.00		10.0		-	-
2	-	0.00		-	–45	-	-
3	353.15	1.10	Silica	50.0		1.458	67.8
4	352.05	–1.10		50.0		-	-
5	-	50.00		-	45	-	-
IMA	∞			0.0		-	-

Prescription 14.2 Kellner eyepiece, comprising production parts from Linos Photonics No. 312315 plano-convex 50 × 22.4 and 322285 doublet 30 × 12.5.
(M = 10XA, TL = 160 mm, OSNA = 0.025, f = 25 mm, FOV = 15 mm)

Surf	Radius	Thickness	Glass	Diameter	Comment	n_d	ν_d
OBJ	∞	–160		15.0		-	-
STO	∞	160.00		8.0	LS	-	-
2	∞	5.89		15.0	Image	-	-
3	∞	4.50	N-BK7	22.4	PC50	1.517	64.2
4	–26.42	19.09		22.4	312315	-	-
5	19.67	3.50	N-SK2	12.5	322285	1.607	56.7
6	–15.51	1.50	SF10	12.5	DB30	1.728	28.4
7	–70.29	16.11		12.5		-	-
IMA	∞			1.8	Exit pupil at eye	-	-

Prescription 14.3 Plössl eyepiece 10X telecentric, comprising production parts from OptoSigma 026-1140 doublets 50 × 30.

(M = 9.4XA, TL = 1000 mm, OSNA = 0.025, f = 25 mm, FOV = 26.6 mm)

Surf	Radius	Thickness	Glass	Diameter	Comment	n_d	ν_d
OBJ	∞	−1000.00		25.0		-	-
STO	∞	1000.00		50.0	LS	-	-
2	∞	16.85		25.0	FS	-	-
3	62.73	3.00	SF2	30.0		1.648	33.8
4	23.40	11.10	BK7	30.0		1.517	64.2
5	−32.50	0.20		30.0	026-1140	-	-
6	32.50	11.10	BK7	30.0	026-1140	1.517	64.2
7	−23.40	3.00	SF2	30.0		1.648	33.8
8	−62.73	15.78		30.0		-	-
IMA	∞			3.2	Exit pupil	-	-

Prescription 14.4 Erfle eyepiece 1923, from U.S. Patent 1,478,704.[54]

(M = 10XA, TL = 1000 mm, OSNA = 0.025, f = 25 mm, FOV = 25.2 mm)

Surf	Radius	Thickness	Glass	Diameter	Comment	n_d	ν_d
OBJ	∞	−200.00		25.0		-	-
STO	∞	200.00		10.0	LS	-	-
2	∞	12.24		25.0	Image	-	-
3	−169.70	1.50	F2	30.0	L11	1.620	36.4
4	25.00	13.80	BK7	36.0	L12	1.517	64.2
5	−34.00	0.20		36.0		-	-
6	65.40	7.00	BK7	36.0	L2	1.517	64.2
7	−65.40	0.20		36.0		-	-
8	28.70	10.70	BK7	32.0	L31	1.517	64.2
9	−28.70	1.50	F2	32.0	L32	1.620	36.4
10	786.00	18.70		28.0	Eye relief	-	-
IMA	∞			2.4	Exit pupil	-	-

Prescription 14.5 Erfle eyepiece 1997. Glass from prior art in U.S. Patent No. 5,691,850 by Arisaka[55] and optimization by Seward.

(M = 10XA, TL = 1000 mm, OSNA = 0.025, f = 25 mm, FOV = 25.0 mm)

Surf	Radius	Thickness	Glass	Diameter	Comment	n_d	ν_d
OBJ	∞	–200.00		25.0		-	-
STO	∞	200.00		35.0	LS	-	-
2	∞	11.48		25.0	Object	-	-
3	–50.50	3.10	SF6	32.0	L11	1.805	25.4
4	37.10	11.60	N-SK16	36.0	L12	1.620	60.3
5	–37.10	0.30		36.0		-	-
6	112.00	6.20	LaK8	36.0	L2	-	-
7	–112.00	0.30		36.0		-	-
8	28.10	12.30	LaK8	32.0	L31	-	-
9	–64.50	3.10	SF6	32.0	L32	-	-
10	417	18.80		28.0	Eye relief	-	-
IMA	∞			2.07	Eye	-	-

Works Consulted

G. R. Fowles, *Introduction to Modern Optics, 2nd Ed.*, Dover Publications, New York (1989).

Matrix methods for reflection and refraction. Reflection coefficients. Maxwell's laws in point format and complex format. Refractive index is derived from conduction of free and bound electrons. Math: algebra, complex numbers.

W. H. Hayt, *Engineering Electromagnetics, 7th Ed.*, McGraw-Hill, New York (2005).

Broad introduction to electromagnetics. Vector operations. Point format of Maxwell's equations.

B. H. Walker, *Optical Engineering Fundamentals, 2nd Ed.*, SPIE Press, Bellingham, Washington (2009).

Numerous descriptions of vision systems with some prescriptions. Several magnifiers and eyepieces. Math: simple algebra.

R. E. Fischer and B. Tadic-Galib, *Optical System Design, 2nd Ed.*, McGraw-Hill, New York (2008).

Practical design considerations. Some discussion of materials.

The American Heritage Dictionary of the English Language, W. Morris, Ed., American Heritage Publishing, Rockville, Maryland (1973).

Origins of words.

Recommended Reading

W. J. Smith, *Modern Optical Engineering, 4th Ed.*, McGraw-Hill, New York (2007).

Introduction to lens design. Paraxial methods. Geometric aberrations. Practical design considerations. Math: algebra, trigonometry.

A. Beiser, *Concepts of Modern Physics*, McGraw-Hill, New York (2002).

Wave propagation. Diffraction. Electron states. Molecular structure. Absorption and emission. Wave-particle duality. Math: algebra, calculus, statistics.

E. Hecht, *Optics 4th Ed.*, Addison-Wesley, Reading, Massachusetts (2001).

Excellent reference for the field of optics. Maxwell's equations in both point and integral format. Refractive index. Wave propagation. Geometric optics. Polarization. Reflection coefficients. Fourier optics. Quantum nature of light. Math: algebra, vectors, calculus.

J. W. Goodman, *Introduction to Fourier Optics 3rd Ed.*, McGraw-Hill, New York (2004).

A classic text book on summation of electromagnetic waveforms. Essential knowledge for an optical engineer. Math: calculus, Fourier methods.

References

1. G. H. Seward, "Two-dimensional space-angle product of a Gaussian beam," *Opt. Eng.* **40**(2), pp. 1959–1962 (2001).
2. B. H. Walker, *Optical Engineering Fundamentals, 2nd Ed.*, SPIE Press, Bellingham, WA (2008).
3. W. J. Smith, *Modern Optical Engineering, 2nd Ed.*, McGraw-Hill, New York (1990).
4. H. R. Phillipp, "Silicon dioxide glass," Fig. 2.4 in F.-T. Lentes, "Refractive index and dispersion," in *The Optical Properties of Glass*, H. Bach and N. Neuroth, Eds., 2nd corrected printing, Springer, New York (1998).
5. E. Hecht, *Optics*, Addison-Wesley, Reading, MA (1987).
6. F. B. Hildebrand, *Advanced Calculus for Applications*, Prentice-Hall, Englewood Cliffs, New Jersey (1976).
7. A. Beiser, *Concepts of Modern Physics*, McGraw-Hill, New York (2003).
8. H. Du, R. A. Fuh, J. Li, A. Corkan, and J. S. Lindsey, "PhotochemCAD: a computer-aided design and research tool in photochemistry," *Photochem. Photobiol.* **68**, pp. 141–142, (1998).
9. R. Y. Tsien, "The green fluorescent protein," *Annu. Rev. Biochem.* **67**, pp. 509–544 (1998).
10. G. H. Seward, "Image contrast of subpixel objects," *Opt. Eng.* **37**(2), 710–716 (1998).
11. E. F. Schubert, *Light-Emitting Diodes*, Cambridge Univ. Press, Cambridge (2006).
12. G. H. Smith, *Camera Lenses*, SPIE Press, Bellingham, Washington (2009).
13. S. Matsui, "Long focal length microlens system," U.S. Patent No. 5,831,775 (1998).
14. G. H. Seward, "Spectroscopy lens for telecentric sensor," U.S. Patent Application 12,333,488 (2008).
15. E. M. Nelson, "Critical microscopy," *J R. Microscop. Soc.*, pp. 282–289 (1910).
16. C. Zeiss, "Description of Professor Abbe's Apertometer," *Proc. Royal Microsc. Soc.* **28**(4), pp. 19–22 (1877).
17. A. Kohler, "Zeitschrift fur wissenschaftl," *Mikroskopie,* **10**, pp. 433–440 (1893),

18. A. Kohler, "A new system of illumination for photomicrographic purposes," *Proc. Royal Microsc. Soc.* **28**(4), pp. 181–185 (1993).
19. ANSI Z136.1, American National Standard for Safe Use of Lasers (2007).
20. T. Marshall and M. Pashley, "LED collimation optics with improved performance and reduced size," U.S. Patent No. 6,547,423 B2 (2003).
21. N. Fukutake, "Microscope objective lens," U.S. Patent No. 6,128,139 (2000).
22. J. Misawa, "Microscope objective lens and microscope," U.S. Patent No. 5,699,196 (1997) .
23. K. Yamaguchi, "Immersion microscope objective lens," U.S. Patent No. 6,519,092 B2 (2003).
24. W. J. Smith, *Modern Lens Engineering, 2nd Ed.*, McGraw-Hill, New York (1992).
25. M. J. Riedl, *Optical Design Fundamentals for Infrared Systems*, SPIE Press, Bellingham, Washington (2001).
26. J. W. Larson, G. H. Seward, G. R. Yantz, D. Johnson, and J. Krogmeier, "Systems and methods for measurement optimization," U.S. Patent No. 7,262,859 B2 (2007).
27. J. R. Krogmeier, I. Schaefer, G. Seward, G. R. Yantz, and J. W. Larson, "An integrated optics microfluidic device for detecting single DNA molecules," *Lab. Chip.* **7**, 1767–1774 (2007).
28. P. Muller, "Optical system," U.S. Patent No. 1,977,689 (1929).
29. R. E. Bitner, "Catadioptrical lens," U.S. Patent No. 2,215,900 (1940).
30. Y. Suehiro,"Reflective type light-emitting diode," U.S. Patent No. 6,641,287 B2 (2003).
31. H. Sakai and T. Kawamura, "Light-emitting diode," U.S. Patent No. 4,698,730 (1987).
32. V. Medvedev and W. A. Parkyn, "Beam-forming lens with internal cavity," U.S. Patent No. 5,575,557 (1998).
33. G. C. Holst and T. S. Lomheim, *CMOS/CCD Sensors and Camera Systems*, SPIE Press, Bellingham, Washington (2007).
34. J. R. Janesick, *Scientific Charge Coupled Devices*, SPIE Press, Bellingham, Washington (2001).
35. Eastman Kodak Company, *KODAK KAI-1020 IMAGE SENSOR* , Revision 5.0 MTD/PS-0205 (2007).
36. Eastman Kodak Company, *KODAK KAC-00401 IMAGE SENSOR* , Revision 1.1 (2009).
37. Eastman Kodak Company, *KODAK PROFESSIONAL T-MAX FILMS,* F-4016 (2007).
38. R. L. De Valois and K. K. De Valois, *Spatial Vision*, Oxford Univ. Press, New York (1990).

39. T. D. Lamb, S. P. Collin, and P. N. Pugh, "Evolution of the vertebrate eye: opsins, photoreceptors, retina and eye cup," *Nature* **2008**(8), pp. 960–975 (2007).
40. J. Sjostrand, N. Conradi, and L. Klaren, "How many ganglion cells are there to a foveal cone?" *Graefe's Arch. Clin. Exp. Ophthalmol.* **232**, pp. 432–437 (1994).
41. P. Artal and R. Navarro, "Monochromatic modulation transfer function of the human eye for different pupil diameters: and analytical expression," *J. Opt. Soc. A* **11**(1), pp. 246–249 (1994).
42. M. Tessier-Lavigne, "Photo transduction and information processing in the retina," in *Principles of Neural Science*, E. R. Kandel, J. H. Schwartz, and T. M. Jessell, Eds., McGraw-Hill, New York (2000).
43. P. Sterling, "Retina," in *The Synaptic Organization of the Brain*, G. M. Shepherd, Ed., Oxford Univ. Press, London (1996).
44. D. H. Kelly, "Visual responses to time-dependent stimuli. I. Amplitude sensitivity measurements," *J. Opt. Soc. A* **51**(4), pp. 422–429 (1961).
45. D. H. Kelly, "Motion and vision. II. Stabilized spatio-temporal threshold surface," *J. Opt. Soc. A* **69**(10), pp. 1340–1349, (1979).
46. R. M. Steinmann,"Effect of target size, luminance, and color on monocular fixation," *J. Opt. Soc. A* **55**(9), pp. 1158–1165 (1965).
47. R. L. DeValois, H. Morgan, and D. M. Snodderly, "Physophysical studies of monkey vision—III. Spatial luminance contrast sensitivity tests of macaque and human observers," *Vis. Res.* **14**, pp. 75–81 (1974).
48. M. S. Banks, A. B. Sekuler, and S. J. Anderson, "Peripheral spatial vision: limits imposed by optics, photoreceptors, and receptor pooling," *J. Opt. Soc. Am.* **8**(11), pp. 1775–1787 (1991).
49. B. B. Boycott and H. Kolb, "The horizontal cells of the Rhesus monkey retina," *J. Comput. Neur.* **148**, pp. 115–140 (1973).
50. R. Jademzik, B. Hladik, and P. Hartmann, "Tailored properties of optical glasses," *Proc. SPIE* **5965**, 59651D (2005).
51. M. Clement, "The chemical composition of glass," Chapter 2.2 in *The optical properties of glass*, H. Bach and N. Neuroth, Eds., Second corrected printing, Springer (1998).
52. SCHOTT North America, Inc., "TIE-29: Refractive index and dispersion," Duryea, Pennsylvania (2007).
53. J. W. Goodman, *Introduction to Fourier Optics*, McGraw-Hill, New York (1968).
54. H. Erfle, "Ocular," U.S. Patent No. 1,478,704 (1923).
55. M. Arisaka, "Eyepiece," U.S. Patent No. 5,691,850 (1997).

Index

George Seward is a consultant in optical design. His instruments are known for easy assembly and consistent performance. His mathematical models define configurations for maximum performance. He developed his expertise in optical design through formal education, practical experience, and continuing education.

His formal education occurred at Tufts University. His bachelors program provided a solid foundation in physics, mathematics, and engineering, while his master's program provided expertise in physical optics and charge transport.

His practical experience is broad. At Innovative Imaging Systems, he developed micro-optics, microscopes, and telescopes. As a consultant, he developed optics for biomedical applications: fluidics, fluorescence, genomics, cytology, and pathology. Many of his optical designs resemble microscopes.

His continuing education developed by necessity. SPIE provided gems of wisdom through seminars, articles, and books. Writing articles and conducting seminars for SPIE extended the depth of his expertise.

His family contains numerous engineers. His father Harold worked at the MIT Instrumentation Lab on guidance systems for the Apollo space program. His brother David developed software at Digital Equipment Corporation. His brother William developed circuitry at IBM. His brother James developed embedded systems at Raytheon. His mother Janet encouraged education in all forms. His great uncle William Seward served as Secretary of State under U.S. presidents Lincoln and Johnson during the abolition of slavery and the purchase of Alaska.